Advanced Manufacturing

Springer

*London
Berlin
Heidelberg
New York
Barcelona
Hong Kong
Milan
Paris
Singapore
Tokyo*

Other titles published in this series:

Intelligent Manufacturing: Programming Environments for CIM
W.A. Gruver and J.C. Boudreaux (Eds)

Automatic Supervision in Manufacturing
M. Szafarczyk (Ed.)

Modern Manufacturing
M.B. Zaremba and B. Prasad (Eds)

Advanced Fixture Design for FMS
A.Y.C. Nee, K. Whybrew and A. Senthil kumar

Intelligent Quality Systems
D.T. Pham and E. Oztemel

Computer-Assisted Management and Control of Manufacturing Systems
S.G. Tzafestas (Ed.)

The Organisation of Integrated Product Development
V. Paashuis

Advances in Manufacturing:: Decision, Control and Information Technology
S.G. Tzafestas (Ed.)

Computer Applications in Near Net-Shape Operations
A.Y,C. Nee, S.K. Ong and Y.G. Wang (Eds)

Parallel Kinematic Machines
C.R. Böer, L. Molinari-Tosatti and K.S. Smith (Eds)

Manufacturing and Supply Systems Management
B. Wu

Dominik Henrich and Heinz Wörn (Eds)

Robot Manipulation of Deformable Objects

With 139 Figures

 Springer

Dominik Henrich, Professor
Embedded Systems and Robotics (RESY), Informatics Faculty, Building 48,
University of Kaiserslautern, D-67653 Kaiserslautern, Germany

Heinz Wörn, Professor
Institut für Prozessrechentechnik, Automation und Robotik (IPR),
Universität Karlsruhe (TH), D-76128 Karlsruhe, Germany

Series Editor

Professor Duc Truong Pham, PhD, DEng, CEng, FIEE
University of Wales Cardiff
School of Engineering, Systems Division,
P.O. Box 917, Cardiff CF2 1XH, UK

ISBN-13: 978-1-4471-1193-1   e-ISBN-13: 978-1-4471-0749-1
DOI: 10.1007/978-1-4471-0749-1

Springer-Verlag London Berlin Heidelberg

British Library Cataloguing in Publication Data
Robot manipulation of deformable objects. - (Advanced
  manufactuirng series)
  1.Robotics 2.Manufacturing processes - Automation
  I.Henrich, Dominik II. Worn, Heinz
  629.8'92

ISBN: 978-1-4471-1193-1

Library of Congress Cataloging-in-Publication Data
A catalog record for this book is available from the Library of Congress

Typesetting: Camera ready by contributors

69/3830-543210 Printed on acid-free paper SPIN 10746690

Table of Contents

Contributors vi

Chapter 1: Introduction 1

   1.1   Motivation 3

   1.2   About this Book 7

Chapter 2: Material Modeling and Simulation 9

   2.1   Energy-Based Modeling of Deformable Linear Objects 11

   2.2   Discrete Element Approach for Non-Rigid Material
        Modeling 29

   2.3   Direct and Inverse Simulation of Deformable Linear
        Objects 43

Chapter 3: Planning and Control Strategies 71

   3.1   Indirect Simultaneous Positioning of Extensible
        Deformable Objects 73

   3.2   A Hybrid Position / Force Approach to the Exploitation
        of Elasticity in Manipulation 91

   3.3   Force- and Vision-Based Detection of Contact State
        Transitions 111

   3.4   Automated Sewing System and Unfolding Fabric 135

Chapter 4: Collaborative Systems 159

   4.1   Manipulation of Sheet Metal by Dual Manipulators 161

   4.2   A Manipulated Deformable Object as an Underactuated
        Mechanical System 175

Chapter 5: Applications and Industrial Experiences 197

   5.1   Simulation of Non-Rigid Materials Handling 199

   5.2   Robotics for Deheading White Fish 211

   5.3   Application of LLW Robots to Distribution Lines 237

   5.4   Flexible Automatic Wiring of Long-Tube Lighting and
        Service Cabinet Modules 255

Contributors

Frank Abegg
Institute for Process Control and
Robotics (IPR)
Computer Science Department
Universität Karlsruhe (TH)
Kaiserstrasse 12
D-76128 Karlsruhe, Germany
Fax: +49-721-608-7141
E-mail: Abegg@ira.uka.de
Http://wwwipr.ira.uka.de/~paro/

Monica Bordegoni
Dipartimento di Ingegneria Industriale
Universita' degli Studi di Parma
Parco Area delle Scienze 181/A
I-43100 Parma, Italia
Fax: +39-0521-905705
E-mail: mb@ied.unipr.it
Http://www.unipr.it/

Rob Buckingham
Oliver Crispin Consulting Ltd
5 Fallodon Way
Henleaze
Bristol BS9 4HR, United Kingdom
Fax: +44-117-983-4397
E-mail: rob@ocrobotics.co.uk
Http://www.ocrobotics.co.uk

Giancarlo Frugoli
Dipartimento di Ingegneria Industriale
Universita' degli Studi di Parma
Parco Area delle Scienze 181/A
I-43100 Parma, Italia
Fax: +39-0521-905705
E-mail: frugoli@ied.eng.unipr.it
Http://www.unipr.it/

Andrea Galimberti
Dipartimento di Ingegneria Industriale
Universita' degli Studi di Parma
Parco Area delle Scienze 181/A
I-43100 Parma, Italia
Fax: +39-0521-905705
E-mail: galimba@ied.eng.unipr.it
Http://www.unipr.it/

Dominik Henrich
Embedded Systems and Robotics (RESY)
Faculty of Informatics
University of Kaiserslautern
Building 48
D-67653 Kaiserslautern, Germany
Fax: +49-631-205-2649
E-mail: henrich@informatik.uni-kl.de
Http: //resy.informatik.uni-kl.de/

Shinichi Hirai
Department of Robotics
Ritsumeikan University
Kusatsu
Shiga 525-8577, Japan
Fax: +81-77-561-2665
E-mail: hirai@se.ritsumei.ac.jp
Http://www.ritsumei.ac.jp/se/~hirai/
index-e.html

Karl-Friedrich Kämper
BJB GmbH & Co.KG
Werler Strasse 1
D-59755 Arnsberg, Germany
Fax: +49-2932-982-390
E-mail: marika.pitts@bjb.de
Http://www.bjb.de/

Kazuhiro Kosuge
Department of Machine Intelligence
and Systems Engineering
Tohuku University
Aoba-yama 01, Aramaki, Aobaku
Sendai 980-8579, Japan
Fax: +81-22-217-6915
E-mail: kosuge@irs.mech.tohoku.ac.jp
Http://www.irs.mech.tohoku.ac.jp/

Kostas J. Kyriakopoulos
Control Systems Laboratory
Mechanical Engineering Department
National Technical University of
Athens
9 Herroon Polytechniou Str.,
Zografou Athens 15700, Greece
Fax: +30-1-772-3657
E-mail: kkyria@central.ntua.gr

Yoshinaga Maruyama
Techniques Development of the
Distribution Department
Kyushu Electric Power Co.
2-1-82, Watanabe-dori Chuo-ku
Fukuoka 810-8720, Japan
Fax: +81 92 712 5236
E-mail:
Yoshinaga_Maruyama@kyuden.co.jp
Http://www.kyuden.co.jp/Distribution/E
nglish/Index2-Distribution.html

Brenan J. McCarragher
Department of Engineering
Faculty of Engineering and Information
Technology
Australian National University
Canberra, ACT 0200, Australia
Fax: + 61-2-6249-0506
E-mail:
Brenan.McCarragher@anu.edu.au
Http://spigot.anu.edu.au/people/brenan/
home.html

Eiichi Ono
Intelligent Systems Division
Electrotechnical Laboratory AIST,
MITI
1-1-4, Umezono
Tsukuba, Ibaraki, 305-8568, Japan
Fax: +81-298-54-5971
E-mail: eono@etl.go.jp
Http://www.etl.go.jp/~eono/

Axel Remde
Institute for Process Control and
Robotics (IPR)
Computer Science Department
Universität Karlsruhe (TH)
Kaiserstrasse 12
D-76128 Karlsruhe, Germany
Fax: +49-721-608-7141
E-mail: Remde@ira.uka.de
Http://wwwipr.ira.uka.de/~remde/

Caterina Rizzi
Dipartimento di Ingegneria Industriale
Universita' degli Studi di Parma
Parco Area delle Scienze 181/A
I-43100 Parma, Italia
Fax: +39-0521-905705
E-mail: rizzi@ied.eng.unipr.it
Http://www.unipr.it/

Herbert Tanner
Control Systems Laboratory
Mechanical Engineering Department
National Technical University of
Athens
9 Herroon Polytechniou Str.,
Zografou Athens 15700, Greece
Fax: +30-1-772-3657
E-mail: htanner@central.ntua.gr

Takahiro Wada
Arimoto and Kawamura Laboratory
Department of Robotics
Faculty of Science and Engineering
Ritsumeikan University
1-1-1 Noji-higashi, Kusatsu
Shiga, 525-8577, Japan
Fax: +81-77-561-2811
E-mail: wachan@robot.club.ne.jp
Http://www.geocities.co.jp/Berkeley-
Labo/1913/index-e.html

Heinz Wörn
Institute for Process Control and
Robotics (IPR)
Computer Science Department
Universität Karlsruhe (TH)

Kaiserstrasse 12
D-76128 Karlsruhe, Germany
Fax: +49-721-608-7141
E-mail: Woern@ira.uka.de
Http://wwwipr.ira.uka.de/

Hidehiro Yoshida
Production Engineering Laboratory
Corporate Production Engineering
Division
Matsushita Electric Industrial Co., Ltd
2-7, Matsuba-cho
Kadoma, Osaka, 571-8502, Japan
Fax: +81-6-6905-4518
E-mail: hyoshida@labo.ped.mei.co.jp
Http://www.labo/pel

Chapter 1
Introduction

Section 1.1

Motivation

H. Wörn

Besides the usage of industrial robots in classical fields of application, e.g., welding, painting and handling, applying robotic systems in additional areas like assembly, machining and measuring is gaining more and more importance. Many research and development projects aiming at the exploitation of these new application areas have been carried out and their results reflect in innovative products and visionary future perspectives.

Under this framework of innovative robot applications, many interesting and challenging topics are related to the handling and processing of deformable objects[1]. By observing the manual processing of such objects, we conclude that a successful task execution especially relies on the following components:

- visual and tactile information,
- multi-fingered hand,
- two-armed corporation,
- prior knowledge and experience.

Developing an ideal and versatile robot system for the handling of flexible objects therefore requires to address the following topics:

- sensor data processing and multi-sensor fusion,
- development of robust and dexterous grippers,
- multi-robot corporation,
- modeling and simulation,

---

[1] In this context, an object is called "deformable" if its deformation must be considered for the manipulation process.

- artificial intelligence.

All of these topics are current research topics, causing the robot handling of flexible materials to be a very challenging field from an academic and scientific point of view.

Since flexible materials are found in almost every industrial product, automated handling of such materials becomes significant from an industrial and economic point of view, too. Two examples can demonstrate this importance: The complete textile industry deals with the processing of highly deformable objects. The price for cloths is mainly depends on the costs of handling and processing the textiles. Though many common tasks like the lapping, (un)folding and positioning have been addressed in many robotic research works and a large variety of prototype systems has been set up, the large-scaled application of robotic handling systems has not yet been achieved. However, the problem of automated textile handling is still on the agenda and the work is being continued, as reflected in different contributions of this book.

The second example is the world of the automotive industry, which is well-known to be one of the most important branches in most industrial countries. Among the large variety of product components to be processed and assembled, many are partly or highly flexible. Though the level of automation and robot implementation is very high in the production, the number of robots is, though increasing, rather small in the final assembly. This causes the assembly costs to be one of the most dominant factors in the final product price.

Because of the large variety of tasks and materials and the automation-friendly environment, the automotive industry is an almost ideal branch to develop, investigate and implement new robot applications. Accordingly, many assembly tasks involving flexible materials, such as, the assembly of hoses, O-ring seals and rubber sealing profiles have been addressed, resulting in robotic prototype systems. For the future, a significant increase of robot applications in the handling of flexible materials can be expected, driven by the high general effort for automating assembly operations in this branch of industry.

When thinking about automated handling, there are two principal ways in dealing with the non-constant object shape. One way is to develop a highly sophisticated, special-purposed robot end-effectors, reflecting the mechanical workpiece properties. Here, most of the effort is required for the hardware design, while using established traditional methods for programming the robot. This approach is proven to meet both key requirements of short cycle time and high reliability.

The other way is to use a more human-like approach, i.e., to keep the design of the hardware as general and flexible (but not necessarily simple and cheap) as possible and to spend more effort for aspects as sensors, sensor data processing, simulation and prediction. The major advantage of this approach is found in the high flexibility and in the possibility of adopting the system to varying tasks, as demonstrated in the human behavior.

Although the flexibility and generality displayed in this second approach are very attractive, it requires to cope with several difficult issues given above, from sensor data fusion to artificial intelligence. The problems related to these topics could not yet been satisfying solved for many tasks less complicated than the handling of flexible materials. In opposite to this, the development of special-purposed hardware has been industrial everyday work for decades. Accordingly, experience and knowledge are widely available here, causing the application of dedicated hardware solutions to be more feasible from a practical point of view.

As the degree of automation in the assembly increases, the number of robot applications in the handling of flexible materials can also be expected to grow. While the first step is the large-scaled integration of special-purpose end-effectors and other hardware tools, the increasing number of complex sensor applications (caused by the demand for flexible and fault-tolerant robotic systems) and the general increase of scientific knowledge will stimulate the introduction of sophisticated and flexible handling systems. This kind of development, starting with hardware solutions dedicated to the economic solution of well-defined problems and leading to more flexible and general systems, can be found in many variations in the industrial history. Only one of many examples is the machine-tools development, starting with simple milling machines and lathes over highly specialized one-purpose equipment to the flexible NC-technology.

One should note that finding a technically and economically significant solution to industrial problems is not a simple one-dimensional problem, but depends on many factors, including technological, social and political aspects, and change in any of them may affect the answer to the initial question as well. Therefore it is wise not to focus on one approach only that seems to be straight-forward, but to investigate a larger variety of principles and methods, even if the industrial application seems to be somewhat futuristic in some cases.

For the automated handling of flexible materials, the aim of this book is to give a survey of both the state of the art and technology. Some of the contributions in this book describe innovative applications of robot systems which could already be established or are expected to have a break-through within the next few years. Others deal with the handling of flexible materials from a more fundamental point of view, highlighting milestones which could be achieved and problems which are under research and will have to be solved in the framework of a long-term perspective.

Section 1.2

About this Book

D. Henrich

Besides the work in the field of manipulating *rigid* objects, currently, there are several research and development activities going on in the field of manipulating *non-rigid* or deformable objects. Several papers have been published on international conferences in this field from various projects and countries. But there has been no comprehensive work which provides both a representative overview of the state of the art and identifies the important aspects in this field.

Thus, we collected these activities and invited the corresponding working groups to present an overview of their research. Altogether, nineteen authors coming from Japan, Germany, Italy, Greece, United Kingdom, and Australia contributed to this book. Their research work covers all the different aspects that occur when manipulating deformable objects. The contributions can be characterized and grouped by the following four aspects:

- object modeling and simulation,
- planning and control strategies,
- collaborative systems, and
- applications and industrial experiences.

In the following, we give a short motivation and overview of the single chapters of the book.

The simulation of deformable objects is one way to approach the problem of manipulating these objects by robots. Based on a physical model of the object and the occurring constraints, the resulting object shape is calculated. In Chapter 2, Hirai presents an energy-based approach, where the internal energy under the geometric constraints is minimized. Frugoli et al. introduce a force-based approach, where the forces between discrete particles are minimized meeting given constraints. Finally, Remde and Henrich extend the energy-based approach to plastic deformation and give a solution of the inverse simulation problem.

Even if the object behavior is predicted by simulation, there is still the question of how to control the robot during a single manipulation operation. An additional question is how to retrieve an overall plan for the concatenated manipulation operations. In Chapter 3, Wada investigates the control problems when positioning multiple points of a planar deformable object. McCarrager proposes a control scheme exploiting the flexibility, rather than minimizing it. Abegg et al. use a simple contact state model to describe typical assembly tasks and to derive robust manipulation primitives. Finally, Ono presents an automatic sewing system and suggests a strategy for unfolding fabric.

In several manipulation tasks, it is reasonable to apply more than one robot. Especially in cases, where the deformable object has to take a specific shape. Since the robots working at the same object are influencing each other, different control algorithms have to be introduced. In Chapter 4, Yoshida and Kosuge investigates this problem for the task of bending a sheet of metal and exploits the relation ship between the static object deformation and the bending moments. Tanner and Kyriakopoulos regard the deformable object as underactuated mechanical system and make use of the existence of non-holonomic constraints. Both approaches model the deformable object as finite elements.

All of the above aspects have their counterpart in different applications and industrial experiences. In Chapter 5, Rizzi et al. present test cases and applications of their approach to simulate the manipulation of fabric, wires, cables, and soft bags. Buckingham and Graham give an overview of two European projects processing white fish including locating, gripping, and deheading the fish. Maruyama outlines the three development phases of a robot system for performing outage-free maintenance of live-line power supply in Japan. Finally, Kämper presents the development of a flexible automatic cabling unit for the wiring of long-tube lighting with plug components.

Chapter 2

Material Modeling and Simulation

Section 2.1

Energy-Based Modeling of Deformable Linear Objects

S. Hirai

Abstract   An energy-based approach to the modeling of deformable linear objects is presented. Many manipulative operations in manufacture deal with deformable linear objects such as wires, cords, and threads with bend, twist, and extension in 3D space. Simulation of their behavior is required in product design and in evaluation of manipulative operations. It is, however, difficult to build a model of deformable linear objects and to simulate their behavior. I will develop an energy-based modeling of deformable linear objects and will show the computation results of their deformation.

First, I will introduce a differential geometry coordinate system, which is appropriate to describing the deformation of linear objects. Second, internal energy of a deformable linear object and geometric constraints imposed on it are formulated. Deformation of a linear object can be computed by minimizing its internal energy under the geometric constraints. This conditional variation problem is solved by parametrization and nonlinear programming. Finally, computational results and experimental results demonstrate the effectiveness of the energy-based approach.

1   Introduction

Recently, many researchers and engineers are interested in the manipulation of deformable soft objects. One serious problem in the manipulation of deformable objects is the modeling issue. Recall that, in the past decades, solid modeling techniques, which provide a systematic method to build a model of a rigid object, have been applied to the rigid object manipulation and many fruitful results have been obtained based on the solid modeling techniques. Unfortunately, few systematic and effective representations of deformable objects are applicable to deformable object manipulation. It should be, therefore, a challenging issue to establish a systematic and coherent modeling technique for deformable objects, which is applicable to their manipulation.

Deformable objects can be divided into three categorizes: 1) linear objects, 2) thin objects, and 3) lump objects composed of limp materials. Many manipulative operations in manufacturing deal with deformable linear objects such as wires, cords, and threads with bend, twist, and extension in 3D space. Electric lines and network cables should be manipulated in the development and in the maintenance of communication systems. Simulating the behavior of deformable linear objects is required in the design of mechanical products and electric products as well as in the task planning of manipulative operations that handle deformable objects. It is, however, difficult to build a model of deformable linear objects and to simulate their behavior.

In this article, I will present a systematic approach to the modeling of deformable linear objects. First, I will introduce a differential geometry coordinate system, which is appropriate to describing the linear object deformation. Second, internal energy of a deformable linear object and geometric constraints imposed on it are formulated using differential geometry coordinates. Deformation of a linear object can be computed by minimizing its internal energy under the geometric constraints. This conditional variation problem is solved by parametrization and nonlinear programming. Finally, computational results and experimental results demonstrate the effectiveness of the energy-based approach.

Related Works

Solid mechanics has been studied for a long time to analyze the deformation of solid bodies [1]. Solid mechanics basically focus on the small deformation of uniform solid bodies. Consequently, it is difficult to formulate and to simulate the large deformation of deformable linear objects, which often show non-uniform properties. A deformed shape of a thread suspended by two points has been analyzed in the calculus of variations [2]. Computing a object shape that minimizes its gravitational energy, it has been found that the shape is described by a catenary [3]. In this analysis, the effect of bend rigidity is out of formulation. Moreover, 3D deformation with twist and extension is out of focus. In computer graphics, deformed shape of cloths has been studied [4]. This research has utilized catenaries to describe the shape of clothes, which inherits the above shortcomings. Modeling of elastic objects has been proposed in computer graphics as well [5]. This research has focused on lump objects composed of elastic materials; linear object deformation is out of scope. In differential geometry, shapes of curved lines in 2D space or in 3D space have been analyzed and a method to describe curved lines has been proposed [6]. Differential geometry focuses on the shape of a curved line and no physical

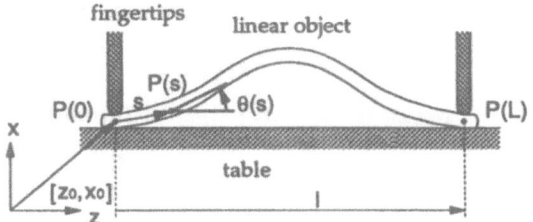

Figure 1: Planar bend deformation of linear object

properties are investigated. Moreover, the twisting of a linear object cannot be described in mathematical differential geometry.

2   Illustrative Example

Let me show an illustrative example to explain the basic concept of the energy-based approach. Consider a linear object on a horizontal table is deformed in a vertical plain, as illustrated in Figure 1. Let L be the length of the object, s be the distance from one endpoint of the object, and $\theta(s)$ be the angle from the horizon at point $P(s)$. Coordinates z and x at point $P(s)$ are then described as follows, respectively:

$$z(s) = \int_0^s \cos\theta ds + x_0, \quad x(s) = \int_0^s \sin\theta ds + z_0,$$

where $[z_0, x_0]^T$ denote coordinates at the left endpoint of the object. Assume that potential energy of the object U is given by the sum of flexural energy U_{flex} and gravitational energy U_{grav}. Namely,

$$U = U_{flex} + U_{grav}. \tag{1}$$

Assuming that bend moment at point $P(s)$ on the object is proportional to the curvature at that point, flexural energy is given by

$$U_{flex} = \int_0^L \frac{1}{2} R_f \left(\frac{d\theta}{ds}\right)^2 ds,$$

where R_f denotes the bend rigidity at point $P(s)$. Gravitational energy is describe by

$$U_{grav} = \int_0^L Dx ds,$$

where D denotes weight per unit length at point $P(s)$.

Due to the interaction between a linear object and other objects such as fingertips and a table, geometric constraints are imposed on the object. Contacts between the object and the fingertips yield some geometric constraints.

Let θ_0 and θ_L be the angles from the horizon at both end points, respectively. Then, we have

$$\theta(0) = \theta_0, \quad \theta(L) = \theta_L. \tag{2}$$

Let l be the distance between the end points of the deformed object. Since displacement vector between the two endpoints is given by $[l, 0]^T$, we have the following equations:

$$z(L) - z(0) = l, \quad x(L) - x(0) = 0. \tag{3}$$

Contact between the object and the table yields other geometric constraints. Note that any point on the object must be located over the horizontal table or on it. This condition is described as follows:

$$x(s) \geq x_0, \quad \forall s \in [0, L]. \tag{4}$$

Computing function $\theta(s)$ that minimizes potential energy U described in eq.(1) under geometric conditions given by eqs.(2), (3), and (4), we can obtain the planar deformation of a linear object. Namely, computing object deformation results in the optimization of energy under geometric constraints. This approach is referred to as *energy-based modeling*.

3 Description of Linear Object Deformation

3.1 Differential Geometry Coordinates

In this section, I will introduce a differential geometry coordinate system, which is appropriate to describe the deformation of linear objects. Let L be the length of a linear object and s be the distance from its one endpoint along the object, as illustrated in Figure 2. In order to describe the deformed shape of the object, I will introduce a coordinate system fixed on space; $O - xyz$. Let $\boldsymbol{x}(s) = [x(s), y(s), z(s)]^T$ be spatial coordinates corresponding to a point $P(s)$ on the object.

Let us investigate how to describe the deformation of a linear object by ignoring its extension. Consider a small part between two neighboring points; $P(s)$ and $P(s+ds)$. The length of this small part is equal to ds in the natural shape and is equal to $\|d\boldsymbol{x}/ds\| \, ds$ in the deformed shape. Thus, the magnitude of the derivative of $\boldsymbol{x}(s)$ with respect to s must be equal to 1, that is, $\|d\boldsymbol{x}/ds\| = 1$ must be satisfied when a linear object has no extension. Unfortunately, parametric representations such as Bésier curves and spline curves, which are commonly used in computer graphics, do not always satisfy this equation. This implies that these representations have a capability of describing the deformed shape but are not appropriate for describing the deformation of a linear object. In order to describe the deformation of a linear object, the relationship between its natural shape and its deformed shape should be represented. From the above discussion, I have found that a new description should be introduced to describe the deformation of a linear object.

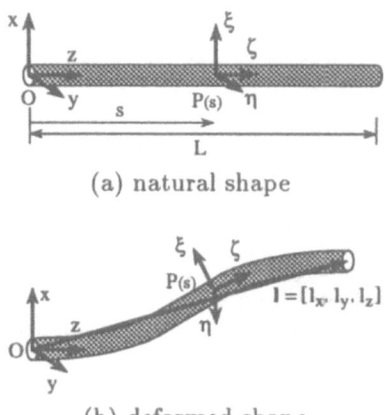

(a) natural shape

(b) deformed shape

Figure 2: Coordinate systems that describe relationship between natural shape and deformed shape

In order to describe the deformation of a linear object, I will introduce a local object coordinates, say $P - \xi\eta\zeta$, at individual points on the object, as shown in Figure 2. Determine the local coordinate system so that ζ-axis is along the linear object at each point. In addition, select the direction of coordinates so that ξ-axis, η-axis, and ζ-axis are parallel to x-axis, y-axis, and z-axis, respectively, in the natural shape. Deformation of a linear object is then given by the relationship between the local coordinate system at each point and the global coordinate system. Let us describe the orientation of the local coordinate system with respect to the space coordinate system using Eulerian angles, $\phi(s)$, $\theta(s)$, and $\psi(s)$. The rotational transformation from $P - \xi\eta\zeta$ to $O - xyz$ is expressed by the following rotational matrix:

$$
\begin{bmatrix} C_\phi & -S_\phi & 0 \\ S_\phi & C_\phi & 0 \\ 0 & 0 & 1 \end{bmatrix} \begin{bmatrix} C_\theta & 0 & S_\theta \\ 0 & 1 & 0 \\ -S_\theta & 0 & C_\theta \end{bmatrix} \begin{bmatrix} C_\psi & -S_\psi & 0 \\ S_\psi & C_\psi & 0 \\ 0 & 0 & 1 \end{bmatrix}
$$

$$
= \begin{bmatrix} C_\phi C_\theta C_\psi - S_\phi S_\psi & -C_\phi C_\theta S_\psi - S_\phi C_\psi & C_\phi S_\theta \\ S_\phi C_\theta C_\psi + C_\phi S_\psi & -S_\phi C_\theta S_\psi + C_\phi C_\psi & S_\psi S_\theta \\ -S_\theta C_\psi & S_\theta S_\psi & C_\theta \end{bmatrix}.
$$

For the sake of simplicity, $\cos\phi$ and $\sin\phi$ are abbreviated as C_ϕ and S_ϕ, respectively.

Let $\xi(s)$, $\eta(s)$, and $\zeta(s)$ be unit vectors along ξ-axis, η-axis, and ζ-axis, respectively, at point $P(s)$ on the deformed shape. Vectors $\xi(s)$, $\eta(s)$, and $\zeta(s)$ coincide to the first column, the second column, and the third column of the above rotational matrix, respectively. Note that $\zeta(s)$ is the unit tangential vector at point $P(s)$. Namely, vector $\zeta(s)$ is equal to the derivative $d\boldsymbol{x}/ds$. Then, the spatial coordinates can be computed by integrating the derivative as follows:

$$x(s) = \int_0^s \zeta(s)ds + x_0, \tag{5}$$

where $x_0 = [x_0, y_0, z_0]^T$ denotes the spatial coordinates at endpoint $P(0)$. Note that this representation satisfies $\|dx/ds\| = 1$.

Extension of a linear object can be taken into consideration by introducing a strain at each point $P(s)$. Let $\varepsilon(s)$ be a strain at point $P(s)$ on a linear object along its central axis. A unit vector along ζ-axis in the natural state are transformed into $(1 - \varepsilon)\zeta(s)$ in the deformed shape due to the object deformation. The spatial coordinates are thus computed by integrating $(1 - \varepsilon)\zeta(s)$ instead of $\zeta(s)$ in eq.(5). The spatial coordinates are then given by

$$x(s) = \int_0^s (1 - \varepsilon)\zeta(s)ds + x_0. \tag{6}$$

From the above discussion, we find that the deformation of a linear object can be represented by four variables: Eulerian angles ϕ, θ, and ψ as well as extensional strain ε. Note that each variable depends on parameter s. This representation is referred to as *a differential geometry coordinate system*. Using a differential geometry coordinate system, the deformation of a linear object can be described in a simple and natural manner.

3.2 Bend, Twist, and Extension of Linear Object

The deformations of a linear object consist of bend, twist, and extension. The extension of the object is represented by $\varepsilon(s)$. In this section, I will describe the bend and the twist of a linear object by Eulerian angles $\phi(s)$, $\theta(s)$, and $\psi(s)$. Let $\kappa(s)$ and $\omega(s)$ be the bend and the twist at point $P(s)$, respectively.

Recall that vector $\zeta(s)$ is the unit tangential vector at point $P(s)$. The bend at point $P(s)$ is thus given by the included angle between $\zeta(s)$ and $\zeta(s + ds)$. Since $\zeta(s)$ and $\zeta(s + ds)$ are unit vectors, the included angle coincides to $\|d\zeta(s)/ds\|$. Thus, it is found that the bend of a linear object at point $P(s)$ is given by

$$\kappa^2 = \left\|\frac{d\zeta}{ds}\right\|^2 = \left(\frac{d\phi}{ds}\right)^2 \sin^2\theta + \left(\frac{d\theta}{ds}\right)^2. \tag{7}$$

Consequently, the bend of a linear object is described by angles $\phi(s)$ and $\theta(s)$.

The twist at point $P(s)$ is defined as the torsional angle around ζ-axis, which causes the difference between $\xi(s)$ and $\xi(s + ds)$. Recall that vector $\xi(s)$ is given by the first column of the rotational matrix. If there exists no bend at point $P(s)$, that is, if vector $\zeta(s)$ is equal to $\zeta(d + ds)$, the torsional angle coincides to the angle between $\xi(s)$ and $\xi(s + ds)$, which is equal to $\|d\xi(s)/ds\|$. Note that derivative $d\xi(s)/ds$ is perpendicular to vector ζ in this case. Considering the bend at point $P(s)$, the torsional angle is given by a component of the derivative perpendicular to ζ-axis. Note that the outer product between $d\xi(s)/d$ and ζ is perpendicular to ζ-axis. In addition, it is found that the magnitude of the outer product is equal to a component of the derivative perpendicular to ζ-axis since ζ is a unit vector. As a result, I find that the component is equal to $\|(d\xi/ds) \times \zeta\|$. The twist of a linear object at point $P(s)$ is thus given by

$$\omega^2 = \left\| \frac{d\xi}{ds} \times \zeta \right\|^2 = \left(\frac{d\phi}{ds} \cos\theta + \frac{d\psi}{ds} \right)^2. \tag{8}$$

Consequently, the twist of a linear object depend on angles $\phi(s)$, $\theta(s)$, and $\psi(s)$.

Assume that a linear object has no twist at any point on the object. Under this assumption, it turns out that angles ϕ and ψ are constant. This implies that the linear object deforms in a plain including z-axis. Without loss of generality, we can assume that the object deforms in $z - x$ plain. Then, angles ϕ and ψ are equal to zero over the object. Consequently, deformation without twist can be described by angle θ and extensional strain ε alone.

4 Energy-Based Modeling

4.1 Internal Energy of Linear Object

In this section, I will introduce a variational principle for statics and will formulate the deformation of a linear object. Variational principles developed in analytical mechanics provide a method to formulate object mechanics as minimization problems [7]. I will formulate the statically stable deformation of a linear object as a minimization problem, which can be solved using computers. Assume that dynamical effects of a linear object is negligible. Let U be the potential energy of the object and W be the work done by external forces applied to the object. The variational principle for statics is given by

$$\delta(U - W) = 0, \tag{9}$$

where δ denotes variational operator. The above equation implies that the internal energy $U - W$ of the object reaches to its minimum at its statically stable shape. In other words, the stable shape can be computed by solving the minimization problem.

Let us first formulate the potential energy of a linear object. Note that the thickness and the width of the object is negligibly small. Applying Bernoulli and Navier's assumption, the potential energy U is described as follows:

$$U = U_{flex} + U_{tor} + U_{ext} + U_{grav}, \tag{10}$$

where U_{flex}, U_{tor}, and U_{ext} represent flexural energy, torsional energy, and extensional energy of the object, respectively, and U_{grav} denotes its gravitational energy.

Assume that bending moment and twisting moment are proportional to the bend angle κ and the twist angle ω at each point, respectively, over the object. Let R_f and R_t represent the flexural rigidity and the torsional rigidity at point $P(s)$, respectively. The flexural energy at point $P(s)$ is given by $(1/2)R_f\kappa^2$ and the torsional energy at the point is given by $(1/2)R_t\omega^2$. The total flexural energy and the total torsional energy are then described as follows:

$$U_{flex} = \int_0^L \frac{1}{2}R_f\kappa^2 ds, \quad U_{tor} = \int_0^L \frac{1}{2}R_t\omega^2 ds.$$

Assume that extensional force is proportional to the external strain at each point over the object. Let R_e denote the extensional rigidity at point $P(s)$. The total extensional energy is then described as follows:

$$U_{ext} = \int_0^L \frac{1}{2}R_e\varepsilon^2 ds.$$

Let D be the weight per unit length at each point of the object. Assume that x-axis is parallel to the gravitational direction. The gravitational energy is then given by

$$U_{grav} = \int_0^L Dx\, ds.$$

Note that quantities R_f, R_t, R_e, and D may vary with respect to variable s.

Next, let us formulate the work done by external forces. Suppose that an external force $\boldsymbol{F}_k$ is applied to the object at an operational point $P(s_k)$, where s_k denotes the distance between the origin and the operational point. Note that coordinates corresponding to $P(s_k)$ at natural shape are given by $\boldsymbol{x}_0(s_k) = [0, 0, s_k]^T$. Thus, the work done by force $\boldsymbol{F}_k$ is described as $\boldsymbol{F}_k^T\{\boldsymbol{x}(s_k) - \boldsymbol{x}_0(s_k)\}$. Assuming that n external forces are applied to the object, the resultant work done by these forces is described as follows:

$$W = \sum_{k=1}^n \boldsymbol{F}_k^T\{\boldsymbol{x}(s_k) - \boldsymbol{x}_0(s_k)\}, \tag{11}$$

where $\boldsymbol{F}_1$ through $\boldsymbol{F}_n$ are external forces acting on the object at point $P(s_1)$ through $P(s_n)$, respectively.

From the above discussion, we find that the internal energy $U - W$ can be described using differential geometry coordinates.

4.2 Geometric Constraints Imposed on Linear Object

Due to the interaction between a linear object and its environment such as fingertips and obstacles, some geometric constraints are imposed on the object. Let us formulate the geometric constraints imposed on the object. The relative position between some points on the object is often controlled during object operations. Consider a constraint that specifies the positional relationship between two points on the object. Let $l = [l_x, l_y, l_z]^T$ be a predetermined vector describing the relative position between two operational points, $P(s_a)$ and $P(s_b)$. The following equational condition must be then satisfied:

$$x(s_b) - x(s_a) = l. \tag{12}$$

The orientation at some points on the object may be controlled during the operation. These orientational constraints are simply described as follows:

$$\phi(s_c) = \phi_c, \quad \theta(s_c) = \theta_c, \quad \psi(s_c) = \psi_c, \tag{13}$$

where ϕ_c, θ_c, and ψ_c are predetermined angles at one operational point $P(s_c)$.

Contact between a linear object and rigid obstacles in operation space also yields other geometric constraints. Note that any point on the object must be located outside or on each obstacle. Let us describe the surface of an obstacle fixed on space by function $h(x) = 0$. Assume that the value of the function is positive inside the obstacle and is negative outside it. The condition that ensures that a linear object is not interfered with by this obstacle is then described as follows:

$$h[x(s)] \le 0, \quad \forall s \in [0, L]. \tag{14}$$

Note that the condition which ensures that an object is not interfered with by obstacles is described by a set of inequalities, since mechanical contacts between the objects constrain the object motion unidirectionally.

From the above discussion, we find that the geometric constraints imposed on a linear object are given by not only equational conditions such as eqs.(12) and (13) but also inequality conditions such as eq.(14). The deformed shape of the object is, therefore, determined by minimizing internal energy $U - W$ under these geometric constraints. Namely, computation of object deformation results in a variational problem under equational and inequality conditions.

4.3 Computational Procedure

Computation of the deformation of a linear object results in a variational problem, as mentioned in the previous section. One method to solve a variational problem is Euler's approach, which is based on the stationary condition in function space. Recall that the geometric constraints resulting from mechanical contacts are unidirectional and are mathematically described by inequalities such as eq.(14). These conditions are nonholonomic constraints [8]. Thus, the shape of an object that minimizes potential energy does not necessarily satisfy the stationary condition. This implies that Euler's approach, which is based on the stationary condition, is not applicable.

Here I will develop a direct method based on parametrization and nonlinear optimization. First, differential geometry coordinates are described by functions of a finite number of parameters. Second, a conditional variational problem with respect to differential geometry coordinates is converted into a conditional optimization problem with respect to the parameters. Then, the converted optimization problem is solved by nonlinear optimization techniques. Let us express differential geometry coordinates $\phi(s)$, $\theta(s)$, $\psi(s)$, and $\varepsilon(s)$ by a linear combination of basic functions $\varphi_1(s)$ through $\varphi_n(s)$. Let a be a set of coefficients in the linear combinations. Substituting the linear combinations into eqs.(10) and (11), internal energy $U - W$ is described by a function of coefficient vector a. The geometric constraints are also described by conditions with respect to the coefficients. In addition, discretizing eq.(14) by dividing interval $[0, L]$ into a finite number of small intervals yields a finite number of conditions. As a result, a set of the geometric constraints is expressed by equations and inequalities with respect to coefficient vector a. The deformation of a linear object is then derived by computing coefficient vector a that minimizes the internal energy under the geometric constraints. This minimization problem under equality and inequality conditions can be solved by use of a nonlinear programming technique such as multiplier method [9]. The deformed shape of the object corresponding to a coefficient vector can be computed using eq.(5). For more details, see [10].

5   Computational Results

In this section, numerical examples are given to demonstrate how the energy-based approach computes the deformation of linear objects. The following set of basic functions is used in the computation:

$$\varphi_1 = 1, \quad \varphi_2 = s, \quad \varphi_{2n+1} = \sin\frac{2n\pi s}{L}, \quad \varphi_{2n+2} = \cos\frac{2n\pi s}{L}. \quad (n = 1, 2, 3, 4)$$

Assume that the length of the object L is equal to 100. Multiplier method and Nelder-Mead method are applied to the nonlinear optimization in the computation.

3D Deformation with Bend and Twist

This example demonstrates the deformation of a linear object with bend and twist. Assume that the object has no extension and that the gravitational energy of the object is negligible. This assumption yields $U = U_{flex} + U_{tor}$. Let us reduce a linear object of its length L along the central axis of the object. Suppose that endpoint $P(0)$ is fixed while the rotation around the central axis of the object at endpoint $P(L)$ alone is allowed. Then, we have the following geometric constraints at the endpoints:

$$\phi(0) = \theta(0) = \psi(0) = 0, \quad \sin\theta(L) = 0, \quad \cos\theta(L) = 1.$$

Assume that dimensionless quantity R_f/R_t, which characterizes the object deformation, is equal to 100. Figure 3 shows the computed deformation corresponding to various values of the distance between two endpoints; $0.8L$, $0.7L$,

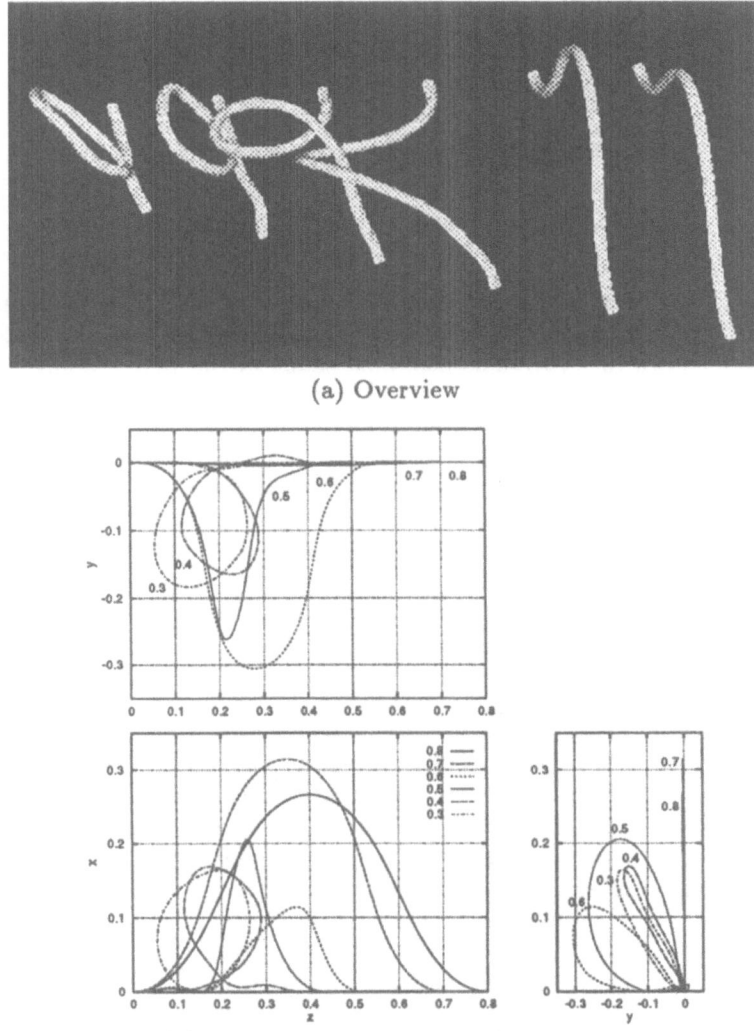

(a) Overview

(b) Top view, front view, and side view

Figure 3: Linear object deformation with bend and twist

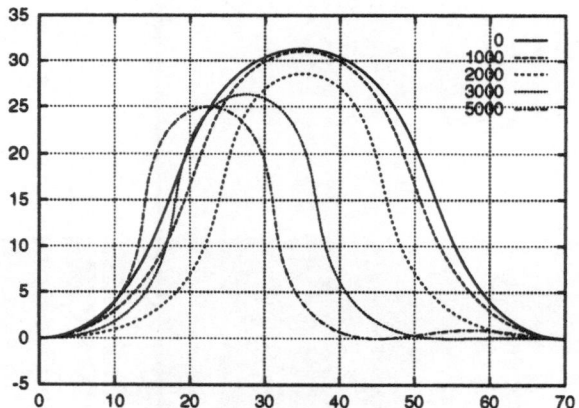

Figure 4: Effect of gravity to linear object deformation

$0.6L$, $0.5L$, $0.4L$, and $0.3L$. Overview of the deformation is illustrated in Figure 3-(a). The top view, the front view, and the side view of the deformed shapes are shown in 3-(b). The deformed shape of the object is involved in x-z plane and is of mode 1 when the distance is equal to $0.8L$ or $0.7L$. The object is twisted and is not involved in any plane when the distance is equal to $0.6L$ or $0.5L$. Namely, the deformation of the object is no longer planar. The object contains one knot, that is, the deformed shape is of mode 2 when the distance is equal to $0.4L$ or $0.3L$. Thus, it turns out that the object shape transits from a knot free shape into a one-knot shape as the distance between the endpoints decreases. Recall that the direction along the central axis of the object is fixed at both endpoints. This implies that the linear object must have a non-planar shape during this transition.

Effect of Gravity

This example demonstrates the effect of gravity. Assume that a linear object has no twist nor extension at any point on the object. This assumption yields $\phi \equiv 0$, $\psi \equiv 0$, and $\varepsilon \equiv 0$ over the object. Potential energy is then given by $U = U_{flex} + U_{grav}$. Normalizing the potential energy by dividing variable s by length L, it turns out that the object deformation is determined by the following dimensionless quantity:

$$\rho = \frac{D}{R_f} L^3.$$

Ratio ρ represents the contribution of the gravitational force to the deformation. Especially, the gravitational force is neglected at $\rho = 0.0$. Impose the following geometric constraints on a linear object; $\theta(0) = \theta(L) = 0$, $z(L) - z(0) = 0.7L$, and $x(L) - x(0) = 0$. Figure 4 shows the deformation of a linear object corresponding to $\rho = 0$, 1000, 2000, 3000, and 5000. As shown in the figure, the height of the object decreases with increasing ratio ρ. In

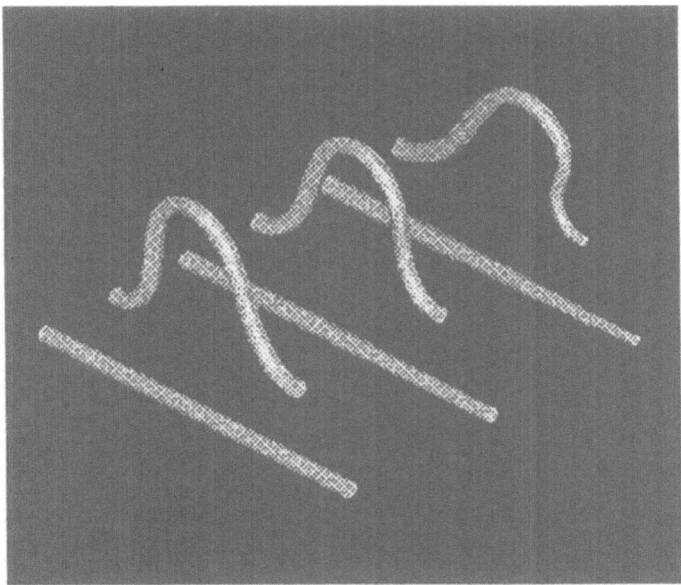

Figure 5: Deformation of non-uniform linear object

addition, the deformed shape is not axial symmetric any more when ρ exceeds 2000. This implies that deformed shapes are unsymmetric when dimensionless quantity ρ exceeds a certain value.

Non-uniform Linear Object

This example demonstrates the deformation of a non-uniform linear object. Recall the flexural rigidity R_f, torsional rigidity R_t, extensional rigidity R_e, and weight per unit length D may depend on variable s. Deformation of a non-uniform object can be computed using the procedure mentioned in Section 4.3, which is based on nonlinear programming techniques. Assume that a linear object has no twist nor extension at any point on the object and that its gravitational energy is negligible. Namely, $U = U_{flex}$. Assume that the flexural rigidity of the object depends on variable s and is given as follows:

$$R_f(s) = 1 + a \cos \frac{\pi s}{L},$$

where a is a constant. The flexural rigidity has its maximum value $1 + a$ at endpoint $P(0)$ and has its minimum value 1 at endpoint $P(L)$. Let us reduce the object so that the following geometric constraints are satisfied; $\theta(0) = \theta(L) = 0$, $z(L) - z(0) = 0.6L$, and $x(L) - x(0) = 0$. The computed deformation is illustrated in Figure 5. The straight lines show the natural shape while the curved lines show the deformed shape. The radius of the object cross section describes the flexural rigidity. The left endpoint corresponds

Figure 6: Kinking of linear object

to $P(0)$ while the right endpoint corresponds to $P(L)$. The front two lines demonstrate the deformation when $a = 0$, that is, when the flexural rigidity is constant. As shown in the figure, the deformed shape is axial symmetric. The middle two lines demonstrate the deformation when $a = 0.5$ and the back two lines demonstrate the deformation when $a = 0.9$. In these cases, the flexural rigidity depend on s and the deformed shape is no longer axial symmetric. The deformed shape inclines towards the right side, where the flexural rigidity is smaller than that at the other side.

Kinking

This example demonstrates the kinking of a linear object. Suppose that a straight rubber band is twisted around its central axis. The deformation of a rubber band will involve knots when the object is twisted enough even if no bend is applied to the endpoints of the object. This deformation is referred to as *kinking*. Let us simulate the kinking of a linear object by energy-based approach.

Assume that the object has no extension and that the gravitational energy of the object is negligible. That is, $U = U_{flex} + U_{tor}$. Let dimensionless quantity R_f/R_t be 1/100. Suppose that endpoint $P(0)$ is fixed while the translational motion along z-axis and the rotational motion around z-axis are allowed at endpoint $P(L)$. Let us twist the object around z-axis at endpoint $P(L)$. Namely, the following constraints are imposed on the two endpoints:

$$\phi(0) = \theta(0) = \psi(0) = 0, \quad \sin\theta(L) = 0, \quad \cos\theta(L) = 1,$$
$$\phi(L) + \psi(L) = \alpha,$$

where α denotes the twist angle around z-axis at endpoint $P(L)$. Figure 6 shows the computed deformation corresponding to various values of twist angle α; 0 to 2π at intervals of 0.2π. The front shape corresponds to $\alpha = 0$ and the back shape corresponds to $\alpha = 2\pi$. The deformation involves twist alone at $\alpha = 0$ and at $\alpha = 0.2\pi$, as shown in the figure. Bend occurs at $\alpha = 0.4\pi$ and the distance between the endpoints is reduced. Since the twist rigidity R_t is 100 times of the flexural rigidity R_f, U_{tor} increases significantly while U_{flex} increases gradually when twist angle α increases. Thus, the bend occurs instead of the twist when twist angle α exceeds a certain value.

6   Experimental Verification

In this section, I will compare the measured deformation and the computed deformation to examine the validity of the energy-based modeling. Note that the proposed method can be applied to the deformation of thin objects around one axis by investigating their cross section perpendicular to the axis. Let us measure the deformation of two sheets of copy paper of $92(\mu m)$ thick shown in Figure 7-(a) and (b), respectively.

Figure 7-(a) shows a rectangle of $200(mm)$ long and $30(mm)$ wide. The bend rigidity R_f and the weight D per unit length of this paper are $10^4(gw \cdot mm^2)$ and $2 \times 10^{-3}(gw/mm)$, respectively. This paper is deformed so that the distance $z(L) - z(0)$ be 180, 140, and $70(mm)$. From the measured deformation, we have estimated that angles $\theta(0)$ and $\theta(L)$ are actually equal to $10°$ and $0°$, respectively. In the computation, we have assumed that angles $\theta(0)$ and $\theta(L)$ are equal to their estimated values. The computed shapes and the measured shapes are plotted in Figure 8-(a). The difference between the computed values and experimental values along vertical axis is 2(mm) at most. Namely, the ratio of the difference to the paper length is 1%.

Figure 7-(b) shows a trapezoid of $200(mm)$ long with a left side $50(mm)$ long and a right side $100(mm)$ long. The bend rigidity R_f and the weight D of this paper are given by $330b(gw \cdot mm^2)$ and $7b \times 10^{-5}(gw/mm)$, respectively, where b denotes the width of the paper. Note that the width b, which is given by $50 + s/4$, depends upon variable s. Thus, the bend rigidity and the weight vary according to variable s. Let us reduce this paper so that the distance $z(L) - z(0)$ is equal to $160(mm)$. The computed deformation and the measured deformation are shown in Figure 8-(b). Estimated values of angles

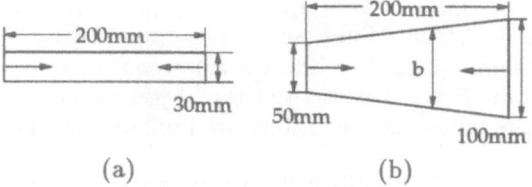

Figure 7: Paper sheets used in experiment

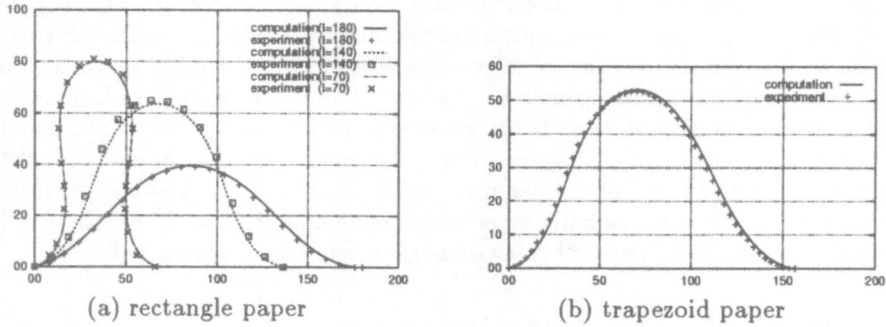

(a) rectangle paper   (b) trapezoid paper

Figure 8:   Comparison between computed deformation
and measured deformation

$\theta(0)$ and $\theta(L)$ are used in the computation. Note that the deformed shape of the object is unsymmetric due to the ununiformity of the bend rigidity and the weight per unit length. The difference between the computed values and experimental values along vertical axis is 2(mm) at most. Namely, the ratio of the difference to the paper length is 1%.

From the above comparison, I have found that the energy-based approach can compute the deformation correctly.

7   Conclusion

I have developed an energy-based modeling of linear object deformation. First, I introduced differential geometry coordinates to describe the deformation of a linear object. I found that the bend, the twist, and the extension of a linear object can be expressed using the differential geometry coordinates. Second, I formulated the internal energy of a linear object and geometric constraints imposed on the object. I found that the computation of linear object deformation results in a variation problem with equational and inequality conditions. Third, I showed some numerical examples to demonstrate the effectiveness of the energy-based modeling. I showed that large deformation of a linear object with bend, twist, and extension of a linear object in 3D space can be computed by the energy-based approach. Deformation of a non-uniform linear object can be computed by this approach as well. Finally, comparison between measured deformation and computed deformation proved the validity of the energy-based modeling of a linear object.

Energy-based modeling of linear objects has been applied to grasping of linear objects [11], insertion of a flexible beam into a hole [12], and modeling of plain knitted fabrics [13]. This approach can be applied to the deformation analysis of electrical cables and wire-driven mechanisms as well as the deformation modeling of vessels, nerves, and chromosomes in biological systems.

References

[1] Fung, Y. C., *Foundations of Solid Mechanics*, Prentice-Hall, 1965

[2] Elsgolc, L. E., *Calculus of Variations*, Pergamon Press, 1961

[3] Irvine, H. M., *Cable Structures*, MIT Press, 1981

[4] Weil, J., *The Synthesis of Cloth Objects*, Computer Graphics, Vol.20, No.4, pp.49–54, 1986

[5] Terzopoulos, D. et al., *Elastically Deformable Models*, Computer Graphics, Vol.21, No.4, pp.205–214, 1987

[6] Auslander, L., *Differential Geometry*, Harper International Edition, 1977

[7] Crandall, S. H., Karnopp, D. C., Kurts, E. F., and Pridmore-Brown, D. C., *Dynamics of Mechanical and Electromechanical Systems*, McGraw-Hill, 1968

[8] Goldstein, H., *Classical Mechanics*, Addison-Wesley, 1980

[9] Avriel, M., *Nonlinear Programming: Analysis and Methods*, Prentice-Hall, 1976

[10] Wakamatsu, H., Hirai, S., and Iwata, K., *Modeling of Linear Objects Considering Bend, Twist, and Extensional Deformations*, Proc. IEEE Int. Conf. on Robotics and Automation, Vol.1, pp.433–438, Nagoya, May, 1995

[11] Wakamatsu, H., Hirai, S., and Iwata, K., *Static Analysis of Deformable Object Grasping Based on Bounded Force Closure*, Proc. IEEE Int. Conf. on Robotics and Automation, Vol.4, pp.3324–3329, Minneapolis, April, 1996

[12] Nakagaki, H., Kitagaki, K., Ogasawara, T., and Tsukune, H., *Study of Deformation and Insertion Tasks of a Flexible Wire*, Proc. IEEE Int. Conf. Robotics and Automation, pp.2397–2402, 1997

[13] Wada, T., Hirai, S., Hirano, T., and Kawamura, S., *Modeling of Plain Knitted Fabrics for Their Deformation Control*, Proc. IEEE Int. Conf. on Robotics and Automation, Vol., pp.1960–1965, Albuquerque, April, 1997

Section 2.2

Discrete Element Approach for Non-Rigid Material Modeling

G. Frugoli, A. Galimberti, C. Rizzi, and M. Bordegoni

Abstract. This paper presents a model adopted to represent and simulate the behaviour of non-rigid material. It is a discrete model, also known as particle-based or mass-spring model. This model describes an object as a set of particles with their mass, radius and other physical properties. The interaction laws among the particles are modelled by means of forces and constraints that determine the dynamic behaviour of the material. Techniques used to control the interaction of the object with its surrounding environment-constraints and collisions are also described in the paper. The same model has been adopted for the haptic rendering of flexible objects. The strategies for supporting real-time interaction with non-rigid objects are discussed as well.

1. Introduction

Today almost all mechanical engineers use CAD systems in the design process. The term *virtual prototyping* means to design and produce mechanical system prototypes on computer. However, the current CAD systems are based on representation schemes that are not adequate to describe non-rigid objects. Since they rely almost exclusively on the geometry of the object, therefore they are oriented to products that are rigid or can be considered as rigid. Since most of the industrial products are flexible, it is necessary to adopt a mathematical model that describes not only the geometry, but also the physical properties of the material: the object's shape depends on the forces applied to it and its initial state.

Techniques and tools for physically-based modelling have been developed and tested. A physically-based model [1] is a mathematical representation of an object, and of its behaviours, that incorporates forces, torques, energies, and other attributes of Newtonian physics. This model is considered as *active*, since it reacts to external forces applied (such as gravity), to constraints or to impenetrable obstacles (such as a table) in ways that an actual object does.

Such a model can further be categorised as follows [2,3]: *continuous* model and *discrete* model. Continuous model is directly based on the elasticity theory [4]; discrete model [5] describes a deformable object by combining very basic mechanical elements whose simulation is carried out by computing the interaction laws among the elements for all types of interactions.

Based on above theories, we have developed a modelling and simulation system [6] for a discrete model, known as *particle-based* or *mass-spring* model. It can be considered as a *macro-molecular* description of the material, for which artificial *forces* acting among the particles are introduced to simulate the inter-molecular attraction and repulsion forces. As a result, we can describe the way the material reacts to external influences and parameters such as elasticity, viscosity and plasticity.

This paper describes the discrete model we have adopted for our prototype and the solutions for various problems related to the dynamic behaviour of the material, e.g. collision detection and response. A system used to model and simulate different types of flexible bodies, e.g. wires, clothes, soft bags, etc, is also discussed.

One of the main issues in physically-based modelling is the computational time required to compute the simulation. Several solutions, including parallel processing, have been proposed to speed up the computation. Some specific applications, such as haptic rendering, require computing the model behaviour in real-time. To overcome the time constraint, we have developed an approximated simulator based on the particle approach, but fast enough for supporting real-time interaction. For this purpose, we need to renounce the level of precision and some of the characteristics that are implemented in the simulator. Renouncing accuracy means firstly to reduce the discretisation level of the model, and then to adopt simplified physics (i.e. omitting computation of friction, damping, self-collisions, etc.) in the simulation of the physical behaviour. This paper presents the approach adopted to develop the approximated real-time simulator and its integration with haptic devices.

2. The Particle-Based Model

The *particle-based* model is the widely known and mostly used discrete model. This model describes an object as a set of particles with their mass, radius and other physical properties. Among two available techniques, *energy based* [7] and *force-based* [5], the first technique is not adequate for dynamic analysis, therefore we consider the force-based approach.

With this approach, the interaction laws among the particles are modelled by forces and constraints that determine the dynamic behaviour of the material. The mathematical representation of the particle system is a system of first-order ordinary differential equations, which can be solved step by step using numerical integration.

In a more rigorous way, given:

$$S = \{P_1, P_2, ..., P_n\} \quad \text{a set of } n \text{ particles}$$

$$m_i \quad \text{the mass of the particle } P_i$$

$$p_i(t), \ v_i(t) \quad \text{position and velocity vectors}$$
$$\text{for } P_i \text{ at time } t$$

the motion equation of the particle P_i is derived from Newton's second law:

$$f_i = m_i \dot{v}_i(t)$$

that can be written as a pair of first order ordinary differential equations:

$$\begin{cases} \dot{p}_i = v_i \\ \dot{v}_i = f_i / m_i \end{cases}$$

Both vectorial functions $p_i(t)$ and $v_i(t)$ can be written as three independent scalar functions (one for each co-ordinate). Therefore, the particle system can be described by a set of $6n$ first order ordinary differential equations.

To determine velocity and position of each particle, we need to solve the equations of the system. The main drawback of this approach is that the system may become stiff. A common solution is to decrease the integration steps, which however increases the computation time.

The best way to cope with the stiffness problem is to use a variable step integration method. By checking on the maximum acceleration applied on the particles, the system can determine the step to use: when accelerations are too high, the step size is reduced, when they are very low, it is increased.

The use of a higher order ODE solver, such as Runge-Kutta is not ideal, because the advantage of having a larger integration step is compromised by the higher computation time required.

2.1 Particles and Forces

The *particles* are the points in which the mass of a body is concentrated. They are the smallest elements of an object. The main attributes of a particle are mass, position, velocity and acceleration, but other physical parameters such as radius, temperature and so on, depending on specific simulation requirements, are also possible.

The particles of an object are related by *forces* that in general represent an interaction law based on the attributes of the particles. Simple forces can be modelled by means of springs and dampers connecting pairs of particles, as shown in Figure 1, while more complex forces are represented by generic mathematical expressions based on the physical parameters of the particles.

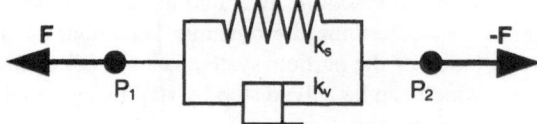

Figure 1. The Kelvin visco-elastic model is used to implement simple forces between pairs of particles

There is a distinction between *internal* and *external* forces. Internal forces represent the internal tension of the material and their purpose is to make the simulated object react as a real one. External forces are used to model the influence of the environment on the object (e.g. gravity and wind) or to impose an artificial force on some points of the object (e.g. compression and traction).

2.2 Representing Objects with Particles and Forces

We have modelled and simulated the behaviour of different types of material: flat objects like fabrics for textile and clothing industry applications, and solid and wire-like objects for applications in different areas.

2.2.1 Flat Objects

A lattice of particles connected by a set of forces, usually represents flat objects such as fabrics, sheets of paper, composite (Figure 2).

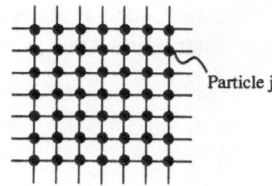

Figure 2. Lattice of particles representing flat objects.

The internal forces used to represent flat material are:
- stretching and repelling;
- bending;
- trellising (shear).

The stretching and repelling forces tend to keep particles at the rest distance. These forces are modelled by means of a Kelvin visco-elastic element between two particles. The bending forces keep the object flat. The trellising forces contrast deformations within the plane.

2.2.2 3D Visco-Elastic Objects

3D visco-elastic objects (e.g. elastomeric parts) are decomposed into small cubes (Figure 3) and forces based on the Kelvin elements as depicted in Figure 4, in order to approximate the isotropic behaviour of the material. Elasticity and viscosity values are determined by using static and dynamic analysis [8].

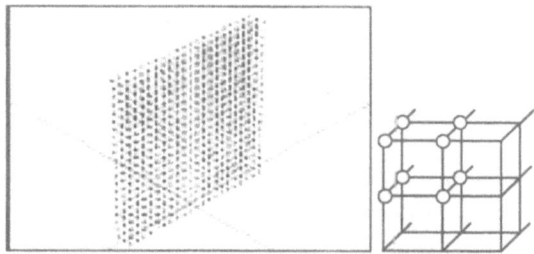

Figure 3. Particle model of 3D visco-elastic object.

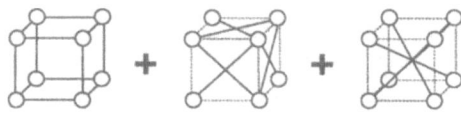

Figure 4. Force distribution for 3D visco-elastic material: along edges + along face diagonals + along internal diagonals.

3. Interaction with the Environment

An object interacts with its environment constantly. These interactions can be characterised as:

- *external forces,* as mentioned above;
- *constraints,* that restrict an object's motions and positions within an allowed range: they must be respected by the object during its motion;
- *collisions with obstacles,* i.e. when a flexible object hits a rigid object, or a flexible object penetrates itself (self-collision). Generally, custom and optimisation techniques are implemented to detect and correct such situations.

We give a description of the techniques adopted to deal with constraints and collisions, as follows.

3.1 Constraints

To fully describe the behaviour of a non-rigid object, it is not sufficient to consider only internal and external forces. It is necessary to take into account other types of interactions between object and its environment, like the presence of rigid obstacles and particular conditions that the object must respect. These conditions are called *constraints*.

A simple example is fixing a point of an object in space. Another example can be the trajectory that some points of an object must follow, as a moving gripper is holding the object itself. In either case, a constraint can be viewed analytically as a mathematical relationship between time and co-ordinates (i.e. points of objects and its environment).

Constraints can be classified into two categories: *equality* and *inequality*. Equality (or *bilateral*) constraints can be represented analytically with an expression of the form $c(p_1, p_2, \ldots, p_n, t) = 0$; they are always active during the simulation, in order to keep the system in a valid state. Inequality (or *unilateral*) constraints are represented by an inequality expression of the form $d(p_1, p_2, \ldots, p_n, t) \geq 0$ and they become active only when the inequality is violated.

There are several methods for applying constraints on a physically-based simulation system using a discrete model. Mostly known techniques are the *penalty method*, the *Lagrangian constraints* and the *dynamic constraints* [9]. We use the dynamic constraints method, since it allows to apply multiple constraints to the same particle and ensures the validity of all the constraints at each step of the simulation. This is done by solving a global linear equation system representing all the active constraints at each time step. The major drawbacks of this approach are the high computational cost and the numerical problems when the linear system becomes ill-conditioned.

In some cases, in order to ensure a constraint is respected, it is sufficient to directly impose a valid state. It is valid to act in this way with simple constraints such as fixed position or fixed trajectory: we can trivially modify the particle position and/or velocity by setting the values that satisfy the constraint at the current time.

Other constraints (like fixed distance between two particles) cannot be maintained so easily, because the valid state is not so obvious. In these cases, the dynamic constraint method solves the problem. When the system tends to go in an unacceptable state, a reaction force is applied to the particles involved, to keep all the constraints imposed.

However, due to numerical approximations, the bilateral constraints are sometimes slightly violated and the system continuously tries to correct this situation. The convergence speed to the acceptable state can be controlled by the parameter τ: if the system is not further perturbed, the valid state is guaranteed to be reached in a maximum time period of $5\,\tau$ seconds.

3.2 Collision Detection

To detect colliding objects, we must find the pairs of entities in the model, whose distance is less than the specified threshold. An extensive search involves checking each particle against each triangle, as well as each triangle side against each other triangle side (Figure 5).

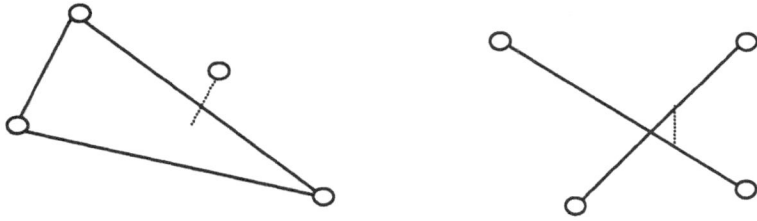

Figure 5. Collision detection: particle/triangle and side/side

The particle/triangle check alone is normally sufficient in almost all the simplest cases, but a more accurate detection requires checking also the sides to avoid compenetration across the edges of the objects.

Some special cases may arise: during a particle/triangle check, the particle may be near a vertex or a triangle side, so we need to solve a particle/particle or a particle/side collision; during a side/side check, one or both sides may collide at one end. Such cases are extremely rare, but it is very important to deal with them, in order to avoid compenetrations.

3.3 Collision Response

The *collision management* system works to prevent particles from crossing an object's surface. It detects if a particle is near a surface (i.e. its distance is less than a threshold value) and changes the state to simulate the effect of the collision. This means that the speeds of all the particles involved are altered to bounce off the colliding object.

For the example in Figure 6, there is a particle (which may be part of a larger object) colliding with a triangle defining a part of the surface of an object. Note that in our system, all the triangles have particles at their vertices.

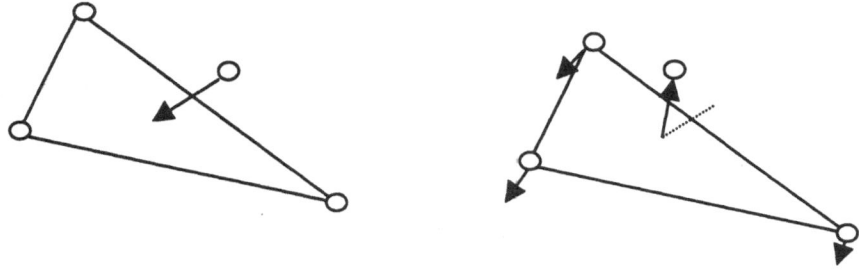

Figure 6. Collision between a particle and a triangle

After the collision, the velocities of the four particles are modified to avoid compenetration: new velocity vectors are computed based on the coefficient of restitution and with respect to conservation of linear momentum. All collisions together form a linear system (usually very sparse), whose solution gives the velocity variations which can be applied to the model.

In a model consisting of n particles, a collision can be seen as an unsatisfied unilateral constraint and written in the form:

$$d_i(p,t) < 0$$

If we have s^* collisions in a particular step, we can write:

$$D^*(p,t) \qquad \text{a vector containing all the collisions}$$

$$J^* = \frac{\partial D^*(p,t)}{\partial p} \qquad \text{a } s^* \times 3n \text{ matrix of partial derivatives}$$

A physically possible solution satisfies the law of conservation of momentum, so we impose:

$$M\Delta v + J^{**} \Lambda = 0$$

Here Δv is the $3n$ sized vector of velocity variations, M is a system inertia matrix including the masses of all the particles, and Λ is a s^* sized auxiliary vector.

It is easy to see that the first term is the momentum of the particles. The J^* matrix describes how the velocity variations must be distributed among the three velocity vector components of each colliding particle. In other words, J^* describes both the direction of the variation vectors and the proportionality among their moduli.

If a particle is not involved in any collision, we can write:

$$M\Delta v = 0$$

Each collision adds a term into the form $J_i^{**} \Lambda_i$, so the particle momentum can change.

The last equation needed imposes the amount of velocity variation depending on the elasticity of the colliding objects. A constraint violation means that $D^*(p,t)$ changes from positive to negative value. To push the system to an acceptable state, we must change a negative $\dot{D}^*(p,t)$ in order for $D^*(p,t)$ to have a positive or null value. If ε is the coefficient of restitution, the last equation will be:

$$\Delta\dot{D}^*(p,t) = -(1+\varepsilon)\dot{D}^*(p,t)$$

Substituting $\dot{D}^*(p,t) = J^*\Delta v$ we obtain:

$$J^*\Delta v = -(1+\varepsilon)\dot{D}^*(p,t)$$

Now we have a linear system, whose solution gives the velocity variations for avoiding compenetration:

$$\begin{bmatrix} M & J^{**} \\ J^* & 0 \end{bmatrix} \begin{bmatrix} \Delta v \\ \Lambda \end{bmatrix} = \begin{bmatrix} 0 \\ -(1+\varepsilon)\dot{D}^*(p,t) \end{bmatrix}$$

Here M is a $3n$ sized diagonal matrix containing the masses of the particles repeated three times (each particle has three equations because it has three components of velocity): $M = diag[m_1, m_1, m_1, m_2, m_2, m_2, \ldots, m_n, m_n, m_n]$.

As already stated, two objects are considered in a collision when the distance is less than a threshold value. We adopted this solution to deal with flat objects such as fabrics, since they have no distinction between inner and outer surface. Setting an appropriate threshold creates a thin virtual volume around the surface; all the particles that enter the volume are detected and their velocities are modified. The threshold value should be small, yet big enough to ensure detection during simulation.

A particular, and sometimes critical, situation in a collision arises when there is a continuous contact, like an object resting on a table. Our system has been tested for this purpose with a good responses both in precision and stability.

3.4 Collision with Rigid Environment and Manipulators

A simulated object interacts with its environment mainly through collisions. A table, a mannequin or a manipulator needs to be modelled but not simulated. There are two possible approaches to rigid objects collisions: integration with particles or *ad hoc* solution.

The integrated approach adopts the same solution technique used for non-rigid objects. All points modeling the rigid environment are considered particles with infinite mass. However, using infinite (or very large) masses increases the stiffness of the solution matrix, which becomes unsolvable. This problem can be resolved by examining how the collision management works, treating an infinite mass particle by setting its J^* values to zero.

This technique is also used for position constrained particles: even if a constrained particle has a finite mass, it cannot change its velocity in response to collisions, therefore acting like a rigid object.

With the other approach, is not necessary to solve the linear system required by the dynamic constraints, thus saving a lot of time. The rigid objects keep fixed position or trajectory so it is very easy to compute directly the correct collision response. In the simplest case a particle collides with a rigid triangle: the only thing to do is reverting the velocity component perpendicular to the surface (for a totally elastic collision). Other cases are slightly more complicated but a similar solution can be found.

3.5 Fast Collision Checking Algorithm

The collision checking is probably the simplest part of a simulator's jobs, however it requires from 70% to 90% of the total computation time. Even for a small model, this search must examine thousands of pairs to find usually a few actual collisions.

To speed up the searching process, we employ the method of *bounding boxes*, a good compromise between simplicity and efficiency. The collision test between two objects can be safely skipped when their bounding boxes are separated.

The method of bounding boxes applied to objects must be improved in some way because a soft object could self-collide. Self collisions occur mainly in fabrics, when they are folded or when they form deep wrinkles. We approach this problem by splitting objects into regions and applying bounding boxes to each region.

An object must be separated into regions in the creation stage, so regions are fixed during the entire simulation. Different regions are considered as different objects; as a result it saves a lot of searching time. Self-collision check inside each single region is always performed: a best approach should perform it only when the curvature of the surface is very high.

We should also pay special attention to partitioning an object. The bounding boxes around adjacent regions are normally overlapped, so a good partition requires at least three regions along the major axes of the object (e.g. a ribbon should be split in three or more subsequent regions, a squared fabric at least into nine regions). The ideal case is to have n regions of n triangles covering an object consisting of n^2 triangles.

It is possible to split a region into sub-regions to form a tree of hierarchical regions and bounding boxes, but experimental evidences suggest to use sub-regions for big objects with thousands of particles.

3.6 Applying Changes to the State

We need to take cautions when we are applying the computed variations to the state of the system. The best time to correct particle's velocity is when the ODE is being solved. A change in velocity needs to be applied *after* the new one is computed, but *before* it is used to update the particle position.

By this way, it guarantees the highest precision during both the modification (because the data to be changed is updated) and the use (because new positions are computed with corrected values). Also, this technique gives the best stability in long simulations and in presence of position constrained particles.

4. A Discrete Model for Haptic Rendering

The model presented above and the prototype we have developed are used to represent and simulate the behaviour of different types of flexible bodies, e.g. wires, clothes, elastomeric parts, soft bags. Some specific application, like haptic rendering, require the computation of the model behaviour in real-time.

We have seen a great demand for haptic interfaces, for example: medical field, design, psychophysics, applications for blind people, telemanipulation and micromanipulation. Some of these applications require the haptic rendering of both rigid and non-rigid objects. The haptic rendering of rigid objects has reached a mature stage. The haptic rendering of non-rigid objects, instead, still has one main issue to be solved, that is the computational time required.

In such a context, we are currently working on the *haptic* rendering of flexible objects by integrating the PHANToM haptic feedback technology [10] into the non-rigid material simulator. The problem is coupling the haptic rendering, which is an inherently real-time process, with the non-rigid material simulation, which is a computational demanding task. A simulation need to run at least a few hundreds times per second to achieve satisfying results for haptic rendering, which is a very high frequency if we want to adopt complete physically-based models and simulate complex objects like body tissues.

The work we are doing is mainly focused on the basic technologies for non-rigid material simulation, but we are also looking for application-specific problems to validate the developed technologies on [11]. The technologies we are studying, or have studied in the past, can be roughly classified according to the source of the haptic data used for rendering.

4.1 Off-line Simulated Data

Our first attempt was to use off-line simulated data in integrating high-fidelity simulator of non-rigid materials with haptic rendering [12]. Our particle-based system offers very good results in terms of fidelity in the simulation of the physical behaviour of the object, however it is too slow to use it directly for haptic rendering.

We observed that every point on some geometrically regular and homogeneous objects behaves in the same way. Therefore, it makes sense to compute off-line behaviour at one point, and use the results run-time at whatever point the actual interaction between the user and the object happens. Examples of such object geometry are the sphere and the halfspace; the assumption was proved by a sample implementation of the halfspace, which yielded good results (Figure 7).

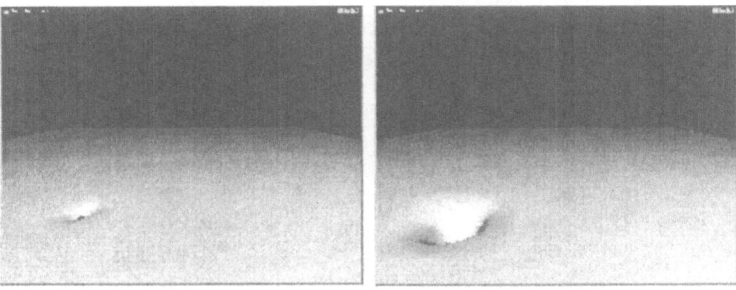

Figure 7. Soft Halfspace deformation simulated using pre-computed data

This approach can be extended to other geometries under some assumptions and approximations, but most geometries cannot be solved by using this approach, especially when the deformations cannot be thought as local or in the case of multi-finger manipulation, where it may be required to off-line pre-compute all the possible position combinations of the different fingers. Although the results of the first experiments were very encouraging, we are not going to further develop this technology, due to the limitations intrinsic of this approach.

4.2 Run-Time Simulated Data

As already stated, a simulation should run at least a few hundreds times per second to achieve satisfying results with haptic rendering; since a trade-off between the speed of a simulation and its accuracy exists, we are currently working on an approximated version of the simulator that is fast enough to supply the haptic data at the required frequency [13]. By adopting simplified physics, we have obtained a faster simulator, which is still based on the physically-based particle model approach and exhibits the following performances:

- the computational time as a function of the input size n, where n is the number of particles, shows to be bound above by $O(n) = n \, log \, (n)$;
- the speed-up on multi-processor computers appears to be equal, or even greater, than the number of CPUs (provided that the size of the model partition which is computed by each processor can fit in its cache memory);
- a 1000-particle model is computed at a rate of about 320 Hz on a 200 MHz Dual P-Pro PC.

In fact, the simulator can use one out of two different computational models: the classical spring-damper-mass model, and a simplified spring-only model. The spring-damper-mass model represents each constraint as a spring-damper system, having separate constants for compression and extension; particle acceleration is derived from the forces exerted on them and their masses. The spring-only model ignores the damping values and the particle masses; updating of particle positions is based on heuristics. The performances reported above are relative to the spring-only model; the spring-damper-mass model shows the same qualitative behaviour, but it is about three times slower. Figure 8 shows a sequence of images showing to the real-time haptic simulation of a piece of foam material.

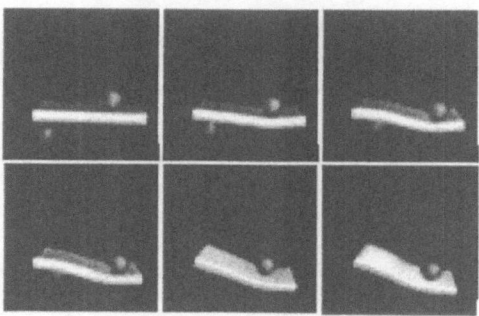

Figure 8. Sequence of pictures showing real-time haptic interaction with a piece of foam material using the approximate model

5. References

1. Barzel R 1992 *Physically-Based Modelling for Computer Graphics: a Structured Approach.* Academic Press

2. Gascuel M P, Puech C 1991 Dynamic Animation of Deformable Objects. In *Eurographics 91 State of the Art Reports*, pp.187-207

3. Ng H N, Grimsdale L 1996 Computer Graphics Techniques for Modelling cloth. *IEEE Computer Graphics and Applications*, 16(5):28-41

4. Terzopoulos D, Platt J, Barr A, Fleischer K 1987 Elastically Deformable Models. *Computer Graphics*, 21(4):205-214

5. Witkin A 1995 Particle System Dynamics. In: *Siggraph 95 Course Notes, Vol 34*

6. Denti P, Dragoni P, Frugoli G, Rizzi C 1996 SoftWorld: A System to Simulate Flexible Products Behaviour in Industrial Applications. In: *European Simulation Simposyum (ESS 96), Vol 2*, pp. 235-239

7. House D H, Breen D E 1998 Representation of Woven Fabrics. In: *Siggraph 98 Course Notes, Vol 31*

8. Colombo G, Frugoli G, Rizzi C 1997 A Computer-Aided System Based on Physically Modelling to Study Flexible Products Behavior. In: *International Conference on Precision Engineering (ICPE 97), Vol 2*, Taipei, pp. 853-857

9. Platt J, Barr A 1988 Constraint Methods for Flexible Models. *Computer Graphics*, 22(4):279-288

10. SensAble Technologies, Inc. *http://www.sensable.com/*

11. Cugini U, De Angelis F 1999 Modelling, Simulation and Haptic Rendering of Nonrigid Objects. In: *MTS*, Rome

12. De Angelis F, Bordegoni M, Frugoli G, Rizzi C 1997 Realistic Haptic Rendering of Flexible Objects. In: *Second PHANToM User Group Workshop*, Cambridge

13. Bordegoni M, Cugini U, De Angelis F 1998 Evolution of Interaction in Physically-based Modelling. In: *6th IFIP WG 5.2 International Workshop on Geometric Modelling: Fundamentals and Applications*

Section 2.3

Direct and Inverse Simulation of Deformable Linear Objects

A. Remde and D. Henrich

Abstract. In this chapter, the quantitative numerical simulation of the behavior of deformable linear objects, such as hoses, wires and leaf springs is studied. We first give a short review of the physical approach and the basic solution principle. Then, we give a more detailed description of some key aspects: We introduce a novel approach concerning dynamics based on an algorithm very similar to the one used for (quasi-) static computation. Then, we look at the plastic workpiece deformation, involving a modified computation algorithm and a special representation of the workpiece shape. Then, we give alternative solutions for two key aspects of the algorithm, and investigate the problem of performing the workpiece simulation efficiently, i.e., with desired precision in a short time. In the end, we introduce the inverse modeling problem which must be solved when the gripper trajectory for a given task shall be generated.

1. Introduction

In this chapter, we consider the quantitative, numerical simulation of the behavior of a deformable, linear object (DLO) handled by a robot manipulator. In addition to the development of special-purpose grippers and the handling based on sensor information, this problem has been addressed in several works. Zheng et al. perform an off-line computation of the gripper trajectory in order to insert a flexible beam into a hole and succeed in performing the task without the additional usage of sensors [1]. Hirai et al. develop an algorithm for the 2D computation of elastically deformable thin parts based on the principle of minimal potential energy [2]. For

DLOs, Wakamatsu et al. extend this approach to 3D-computation [3] and to the consideration of dynamics based on Hamilton's principle [4].

These works demonstrate that simulating the behavior of DLOs numerically is possible. However, in practice, we need to consider some additional items.

- While dynamics needs to be considered in some cases, this is not necessary in many other cases. Therefore, it is desirable to use an algorithm which allows a changing from static to dynamic computation with little additional effort. Yet, this is not possible when directly employing Hamilton's principle.

- In many practical applications, the workpiece is not only deformed elastically, but also plastically. However, the occurrence of plastic deformation is not being considered in the state of the art. Besides the physical effect itself, we find that an appropriate internal representation of the workpiece shape is necessary when considering plastic deformation.

- When performing simulation on a workpiece, it is generally desirable to perform the computation with sufficient precision in a short time, i.e., to do it in an efficient way. Therefore, we consider different alternatives for some key aspects of the computation algorithm, which are of major influence on the computation time. Additionally, we investigate the influence of some basic computation parameters on both computation time and precision of the results.

- Besides the selection of appropriate computation parameters, parallel processing is an obvious way to reduce the computation time. Thus, we investigate different possibilities of parallel computation and discuss the advantages and limits.

- One major application field of the simulation of deformable objects is the off-line generation of gripper trajectories for a given (assembly) task. That is, for each time step of the assembly process, the task defines certain boundary conditions which must be fulfilled by the workpiece shape. The goal is to compute a gripper trajectory that fulfills these boundary conditions. Because this problem is just an inverse to the computation of the workpiece shape for given boundary conditions, we call it the *"Inverse Simulation Problem"*. In the last part of this section, this kind of task is discussed.

2. Principal Approach

2.1 Physical Principle

In this chapter, we give a short review of the physical approach used for the simulation of DLOs as well as the basic computation principle.

According to the fundamental physical principle of minimal potential energy, dynamic systems assume a minimum of their total potential energy W in any stable state. This holds true not only for systems of discrete elements, e.g., lumped masses and springs, but also for a deformable continuum like DLOs. Based on this principle, the shape of a deformable object can be computed rather easily, if the boundary conditions are known. Neglecting linear extension, potential energy due to gravity, bending and torsion must be considered. Thus, the following optimization problem has to be solved.

$$W = \int_0^L (W'_{grav} + W'_{bend} + W'_{tor}) \, ds \rightarrow \min \qquad (1)$$

In Eqn. 1, L is the length of the workpiece, $s \in [0, L]$ is the curve length measured along the workpiece. W'_{grav}, W'_{bend} and W'_{tor} are the potential energy caused by gravity, bending and twisting (per length) respectively. For each point of the workpiece, they are given as follows:

$$W'_{grav}(s) = \rho \, A \, |g| \, z(s) \qquad (2a).$$

Here, ρ and A are the density and the cross section area, g is the acceleration vector due to gravity, and $z(s)$ the coordinate of the workpiece point along g.

Being R_{bend} and R_{tor} the (constant) bending and torsional rigidity, and $\kappa(s)$ and $\tau(s)$ the local curvature and twisting, the respective potentials are

$$W'_{bend} = \frac{1}{2} R_{bend} \kappa(s)^2 \qquad (2b)$$

$$W'_{tor} = \frac{1}{2} R_{tor} \tau(s)^2 \qquad (2c).$$

2.2 Computation

When computing the workpiece shape, the goal is to determine those functions $W'_{grav}(s)$, $W'_{bend}(s)$, and $W'_{tor}(s)$ that fulfill the condition given in Eqn. 1. In order to perform this computation, the following steps are performed:

- First, a vector $q(s)$ is determined which fulfills the following requirements:

  - $q(s)$ describes the workpiece shape (Cartesian coordinates $x(s)$ of each workpiece points) unequivocally

    $$x(s) = f(q(s)).$$

  - The total potential energy $W'(s)$ per length (having the portions W'_{grav}, W'_{bend}, and W'_{twist}) can be expressed as a function of $q(s)$

    $$W'(s) = f(q(s)).$$

- For a three-dimensional computation, $q(s)$ has three components: $q(s) = [q_1(s), q_2(s), q_3(s)]^T$. Thus, Eqn. 1 is transformed into

$$W = \int_0^L f(q(s)) \, ds \rightarrow \min \qquad (3),$$

and determining the object shape means to compute the components $q_i(s)$ of vector $q(s)$ in order to satisfy Eqn. 3. This is a "calculus of variations" problem, described by a set of partial differential equations (Eulerian equations). Because an analytical solution for these equations can not be found in most cases, we use the well-known approximation method introduced by Ritz [5].

- In this method, the single components of $q(s)$ are expanded into a series with N_c terms.

$$q_i(s) = \sum_{j=0}^{N_c-1} c_{i,j}\, Q_j(s) \qquad (4).$$

Here, $Q_j(s)$ are the basis functions of the series and $c_{i,j}$ are the according coefficients. Thus, $q(s)$ is represented by a vector c having $N = 3N_c$ components. By this step, the problem of computing a vector $q(s)$ of functions is reduced to the problem of computing the vector c of coefficients, and Eqn. 1 is finally transformed into

$$W(c) = \int_0^L f(c,s)\, ds \rightarrow \min \qquad (5).$$

- Since this equation can not be solved analytically as well, the integral in Eqn. 5 is computed by numerical integration and the resulting discrete minimization problem is solved by numerical optimization in Multidimensions.

For static computations, this approach is straight forward. In [2, 3], it is used for computing the static shape of DLOs.

3. Consideration of Special Aspects

3.1 Dynamics

When regarding a robot system manipulating a deformable workpiece, the goal is generally not to compute a single workpiece shape with given boundary conditions, rather than to compute the shape of the object at each point of the gripper trajectory. Wakamatsu et al. [4] point out that the resulting object shape is highly sensitive to the velocity of the gripper motion. If this velocity is sufficiently small ($v_{\text{Gripper}} \rightarrow 0$), the workpiece can be regarded as resting in each simulation step. Inertial forces (causing, e.g., workpiece oscillations) are neglected. This behavior is called *quasi-static*.

When the DLO is manipulated fast, the inertia forces caused by the object acceleration can not be neglected. Therefore, the shape in step i depends on the results of the position and velocity in step $i-1$ and the acceleration between step $i-1$ and step i. This behavior is called *dynamic*.

The principle of minimal potential energy (Eqn. 1) holds true only for static or quasi-static computations. An extension towards the consideration of dynamics leads to Hamilton's principle. With T being the kinetic energy, Eqn. 1 is replaced by

$$S = \int_{t_0}^{t_0+\Delta t} L\, dt = \int_{t_0}^{t_0+\Delta t} (T - W)\, dt \overset{!}{=} \min \qquad (6a).$$

A formulation equivalent to Eqn. 6a is given by the Euler-Lagrange equations

$$\frac{d}{dt}\left(\frac{\partial L}{\partial \dot{q}_i}\right) - \frac{\partial L}{\partial q_i} \overset{!}{=} 0 \qquad (6b).[1]$$

Wakamatsu et al. [4] present a method that computes the shape of DLOs dynamically based by solving Eqn. 6b.

Hamilton's principle is a straight forward method when considering dynamics, but it requires a significantly extended approach compared to the static computation. However, since dynamic computation (which is rather time consuming compared to static computations) is not required in many cases, it is not advantageous. Therefore, we propose a novel approach that allows one to perform static and dynamic computations using the same principle.

We first consider an isolated mass element Δm of the workpiece. Given x_i, $i \in [1..3]$ as its Cartesian coordinates, we obtain (because of $\partial T / \partial x_i \equiv 0$ and $\partial W / \partial \dot{x}_i \equiv 0$) the well-know equation for the motion of a mass element in a potential field from Eqn. 6b

$$\Delta m \, \ddot{x}_i = -\frac{\partial W}{\partial x_i} \qquad (7a),$$

or, in vectorial form

$$\Delta m \, \ddot{x} = -\nabla_x W \qquad (7b).[2]$$

With $x(t_0) = x_0$, $\dot{x}(t_0) = v_0$ and $x(t_0 + \Delta t) = x_1$, $\dot{x}(t_0 + \Delta t) = v_1$, the acceleration $\ddot{x}_1$ in the time interval $[t_0 ... t_0 + \Delta t]$ is given by

$$\ddot{x}_1 = \frac{v_1 - v_0}{\Delta t} \qquad (8).$$

Additionally, we find with $\bar{v}_{0,1}$ being the mean velocity in the considered time interval

$$\bar{v}_{0,1} = \frac{x_0 - x_1}{\Delta t} = \frac{v_0 + v_1}{2} \quad \leftrightarrow \quad v_1 = 2\frac{x_1 - x_0}{\Delta t} - v_0.$$

By combining this relation with Eqn. 4, we obtain $\ddot{x}_1$ as

$$\ddot{x}_1 = \frac{2}{\Delta t^2}(x_1 - x_0 - v_0 \Delta t).$$

This relation can be inserted into Eqn. 7b for the motion of Δm in potential field W.

$$\nabla_x W(x_1) + \frac{2\Delta m}{\Delta t^2}(x_1 - x_0 - v_0 \Delta t) = 0 \qquad (9).$$

The left side of this equation is equal to the gradient of

$$U(x) = W(x) + \Delta m\left(\frac{x - x_0 - v_0 \Delta t}{\Delta t}\right)^2 \qquad (10)$$

---

[1] In the (quasi) static case, we find that Eqn. 6b is equivalent to the differential formulation of Eqn. 1, because of $T \equiv 0$ and $\partial W / \partial \dot{q}_i \equiv 0$.

[2] Eqn. 3b is equivalent to Newton's law of motion $F = ma$.

with respect to x. Thus, to solve Eqn. 9, we have to deal with the optimization problem

$$U(x) \to \min \tag{11}.$$

The position x which satisfying Eqn. 11 is the desired position x_1 of the mass element at time $t_0 + \Delta t$. Note that the acceleration in time interval Δt is assumed to be constant here. This is equivalent to $\Delta t \to 0$. Thus, Δt must be sufficiently small for the numerical computation.

As formulated in Eqn. 1, the static position of the DLO can be determined by minimizing its potential energy W. This holds true for a single lumped mass Δm, too. Therefore, computing the position x of a lumped mass dynamically is reduced to solving the minimization problem for $U(x)$ given in Eqn. 11, instead of $W(x)$. The only difference between U and W is the additional term

$$W_{\text{dyn}} = \Delta m \left(\frac{x - x_0 - v_0 \Delta t}{\Delta t} \right)^2 \tag{12}$$

in U. This relation holds true not only for a single lumped mass, but also for a number of N elements moving in a potential field. For considering a continuum, a border crossing $N \to \infty$ must be carried out. Doing this,

$$W_{\text{dyn}} = A\rho \int_0^L \left(\frac{x(s) - x_0(s) - v_0(s)\Delta t}{\Delta t} \right)^2 ds = \int_0^L W'_{\text{dyn}} \, ds$$

is obtained from Eqn. 12. W_{dyn} takes the workpiece dynamics per length into account.[3] Therefore, the position x of each workpiece point at time $t = t_0 + \Delta t$ can be computed from the known position x_0 and velocity v_0 for $t = t_0$ by solving

$$U = \int_0^L (W'_{\text{grav}} + W'_{\text{bend}} + W'_{\text{tor}} + W'_{\text{dyn}}) \, ds \stackrel{!}{=} \min \tag{13}.$$

Switching from quasi-static to dynamic computation can now be done by simply adding the term W'_{dyn} to the integrand.

3.2 Plastic Deformation

3.2.1 Considering Plastic Deformation

So far, we have assumed the workpiece deformation to be totally elastic. However, in many practical applications, the occurrences of considerable plastic deformation are in presence. Because of the workpiece's behavior depends on many factors, the exact consideration is rather difficult, even if only linear stress without bending or twisting is assumed. The most important influence factors are:

- direction of force (tensile or compression load)
- duration of force application

---

[3] Though W'_{dyn} represents the workpiece dynamics in the minimization problem, it is not the kinetic energy of the object!

- velocity of force increase and decrease

Additionally, the behavior is generally different for linear stress, bending and twisting. However, an exact consideration is not necessary in many cases, and a rather coarse approximation is sufficient.[4] Thus, we use the following assumptions for significantly simplifying the problem.

The workpiece behavior is assumed to be *elastic-ideal plastic* [6], i.e., the stress-strain relation is given by Figure 1 (left), causing a deformation behavior according to Figure 1 (right). Starting in the stress-free state, the internal stress $\sigma(\varepsilon)$ increases linearly (Hook's law, Phase 1). Besides a certain yield strain ε_E, the stress remains constant for increasing strain (Phase 2). For a subsequent force relief, the stress decreases on a parallel to the original Hook's straight line (Phase 3) with strain ε_p remaining after complete stress release. This behavior implies especially that the stress-strain relation is independent of the deformation history, which does generally not hold true in reality.

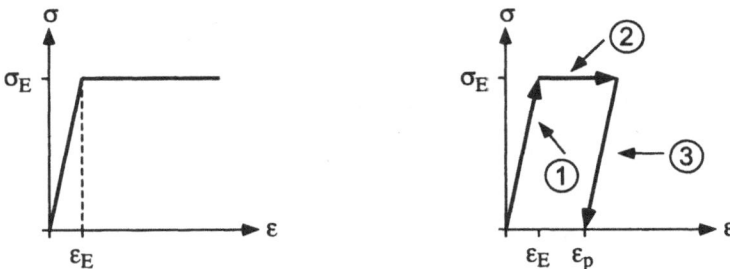

Figure 1: Left: Stress-strain diagram for elastic-perfectly plastic material behavior (σ: stress, ε: strain, σ_E: yield stress, ε_E: yield strain). Right: stress-strain-cycle for tensile load (ε_p: residual strain).

The idealized consideration of plastic deformation according to Figure 1 holds true for a linear (tensile) load. For bending and twisting a similar behavior can be assumed. In these cases, strain ε must be replaced by curvature κ or torsion τ, respectively. For the consideration of bending, the following additional simplification is used.

For bending loads, the amount of strain is different for the single fibres, depending on their distance from the neutral axis. Thus, for a circular workpiece of radius R, the deformation has to be considered separated for all distances $r \in [0, R]$ from the neutral axis. As simplification, we assume the curvature κ to be identical for the total cross section, with plastic deformation occurring for κ being greater

---

[4] An exact consideration often fails because of the missing material parameters.

than the threshold curvature κ_E. This simplification can be derived directly from the assumption of a one-dimensional object of negligible cross-section.[5]

3.2.2 Modification of the Computation Algorithm

The simulation of plastic deformation requires the following modifications in the computation algorithm.

For computing the potential energy due to bending and twisting, the proportional relations $W_{bend} \cong \kappa^2$ and $W_{tor} \cong \tau^2$ (Eqn. 2) are not valid. Instead, the relation

$$W'_{bend} = \int_0^\kappa \sigma(\tilde{\kappa}) d\tilde{\kappa}$$

must be used for computing the bending energy (and, correspondingly, the torsional energy), with $\sigma(\tilde{\kappa})$ given by the stress-strain-diagram.

The occurrence of plastic deformation implies a change in the stress-free workpiece shape. Thus, the stress-free workpiece shape used for the computation in simulation step i is given by the total plastic deformation of the previous simulation steps 0, 1, ..., i-1. The additional deformation computed in step i contains two portions. Its plastic portion (according to the stress-strain-relation) must be added to the plastic deformation computed in the previous steps, its elastic portion of step i is ignored in the subsequent simulation steps.

Thus, the computation algorithm for step i is as follows:

```
1    D_New := 0;
     Optimum := false;
2    repeat
3        D := D_plastic_{i-1} + D_New;
4        W_grav := W_grav(D);
5        W_bend := W_bend(D_New);
6        W_tor := W_tor(D_New);
7        W := W_grav + W_bend + W_tor;
8        if Minimum(W) then
9                Optimum := true
10       else D_New := D_New + ΔD;
11   until Optimum;
12   D_plastic_i := D_plastic_{i-1} + PlasticPortion(D_New);
```

In this algorithm, the vector D represents the workpiece deformation due to bending and twisting[6], and $D\_plastic$ is its plastic portion according to the stress-strain-relation. D_New is the (additional) deformation computed in simulation step i

---

[5] In reality, the different curvature (and thus, bending stress) of the single fibres causes internal stress within the workpiece.

[6] For the adding of deformations, please refer to Section 3.3.2.

and ΔD is the variation of **D_New** in each iteration of the numerical optimization. The optimization itself is represented by the **repeat** ... **until**-loop.

In each step of the numerical optimization, only **D_New** is used for computing the energy due to bending and twisting, while both **D_New** and the plastic deformation of the previous simulation steps are used for computing the potential due to gravity. This is due to the fact that the energy required for plastical bending and twisting is 'lost' irreversibly and, thus, does not have to be considered for the future simulation steps. However, the stress-free workpiece shape (and, thus, the gravity potential) is influenced by the previous plastic deformation.

3.3 Workpiece Representation

3.3.1 Representation by Curvature and Torsion

In line 3 and line 12 of the algorithm given in Section 3.2.2, different deformation portions have to be added. The addition of deformations causes some restrictions to the internal representation of the workpiece shape.

According to Section 2, the workpiece shape is represented by a vector $q(s)$ of three functions q_i which must be suited for computing both the Cartesian position of each workpiece point and the potential energy. However, we did not give any further information on what kind of functions should be used.

One approach is to directly use the Cartesian coordinates $x(s) = [x(s), y(s), z(s)]^T$ of each workpiece point. However, it is found that computing the potential energies due to bending and twisting is rather complicated. Therefore, this representation is not desirable.

Another approach is to describe the accompanying trihedron (i.e., a "local" Cartesian coordinate system consisting of tangent vector $t(s)$ and two normal vectors n, b), $\{t, n, b\}$ for each point $s \in [0, L]$ of the DLO with respect to a global Cartesian coordinate system. With this representation, the global coordinates of each point on the workpiece can be computed easily by just integrating the tangent vector over s, and there exists a rather simple relation between these vectors[7] and the local amount of curvature and torsion. This method is used by Wakamatsu et al. [3, 4], describing the orientation of the accompanying trihedron by three Eulerian angles φ, θ, ψ. Thus, the vector of functions representing the workpiece shape is $q(s) = [\varphi(s), \theta(s), \psi(s)]^T$.

The main drawback of Eulerian angles (or, similarly, roll, pitch and yaw angles) is that curvature and torsion have impact not only on one, but on all of them. Thus, the relation between workpiece deformation (given by curvature and torsion) on the one hand, and Eulerian angels on the other hand, is not invertable. In conclusion, given a set of Eulerian angles (as functions of s), it is not possible to compute curvature and torsion, and given two sets of Eulerian angles, it is not possible to add the corresponding deformations.

---

[7] and their first derivatives with respect to s

This problem will not occur as long as only elastic deformation is regarded. But according to the algorithm given in Section 3.2.1, considering plastic deformation requires the addition of deformations (line 3 and line 12). Thus, a representation by Eulerian angles is not possible.

To avoid this problem, we choose a different approach, in which curvature and torsion are directly used as internal representation of the DLO. Being $\{t(s), n(s), b(s)\}$ the accompanying trihedron at position s, we use the following functions $q_d(s)$, $q_\kappa(s)$ and $q_\tau(s)$ with $q(s) = [\, q_d(s), q_\kappa(s), q_\tau(s)]^T$ to compute the accompanying trihedron at position $s+\Delta s$.

$q_d(s)$ represents the local direction of curvature, while q_κ is the local amount of curvature. In order to describe the object bending from point s to point $s+\Delta s$ on the DLO, the accompanying trihedron is rotated by the (infinitesimal) angle

$$dw_\kappa = q_\kappa(s)\,ds$$

about the axis which is formed by rotating one of the normal vectors by q_d about tangent vector t.[8] Figure 2 (top) shows a DLO section with the accompanying trihedron and the rotation axis. The result of this rotation is a new trihedron $\{b'(s), n'(s), t'(s)\}$, as shown in Figure 2 (bottom).

$q_\tau(s)$ is the local amount of torsion. Torsion is performed by rotating the trihedron $\{b'(s), n'(s), t'(s)\}$ by the angle

$$dw_\tau(s) = q_\tau(s)\,ds$$

about the t'-axis.[9] As result of this second rotation, the accompanying trihedron $\{t(s+ds) = t'(s), n(s+ds), b(s+ds)\}$ at position $s+ds$ is obtained.

3.3.2 Adding Deformations

With the method derived above, adding deformations according to Section 3.2.2 is rather simple. When adding two deformation portions, (I) and (II), each of them has a value for q_d, q_κ, and q_τ.

Because twisting is always performed around the tangent vector (i.e., the rotation vector is identical for both portions), the rotations can be simply added

$$dw_\tau = (q_{\tau,I} + q_{\tau,II})\,ds \qquad (14a).$$

For the curvature, the directions of both portions are generally different. Here, the corresponding rotation vectors $dw_{\kappa,I}$ and $dw_{\kappa,II}$ are added.[10] These vectors lay in the plane given by $n(s)$, $b(s)$, with the absolute values $dw_{\kappa,I}$ and $dw_{\kappa,II}$. The directions of curvature are given by $q_{d,I}$ and $q_{d,II}$. According to Figure 3, direction q_d and absolute value dw_κ of the resulting curvature are given by

$$dw_\kappa^2 = dw_{\kappa,I}^2 + dw_{\kappa,II}^2 + 2\,dw_{\kappa,I}\,dw_{\kappa,II}\cos(q_{d,I} - q_{d,II})$$

$$\tan(q_d) = \frac{dw_{\kappa,I}\sin(q_{d,I}) + dw_{\kappa,II}\sin(q_{d,II})}{dw_{\kappa,I}\cos(q_{d,I}) + dw_{\kappa,II}\cos(q_{d,II})} \qquad (14b).$$

---

[8] Without loss of generality, we assume that the rotation is performed about $n(s)$.

[9] $q_\kappa(s)$ and $q_\tau(s)$ are the local amount of curvature and torsion at position s.

[10] Adding the rotation vectors is permissible because both rotation angles are infinitesimal.

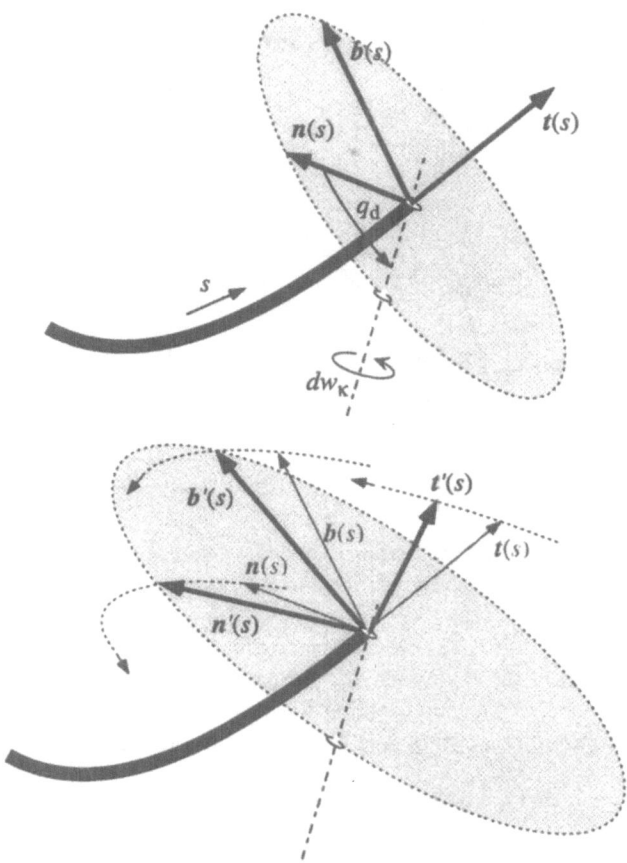

Figure 2: Expression of DLO curvature by its direction q_d, and amount, $dw_\kappa = q_\kappa ds$. Top: Rotation of trihedron $\{t(s), n(s), b(s)\}$. Bottom: Resulting trihedron $\{t'(s), b'(s), n'(s)\}$.

3.3.3 Transformation into Global Coordinates

The vector $q(s)$ given above is the internal representation of the workpiece shape for solving the minimization problem given in Eqn. 1. However, the final goal is to compute the coordinates of each point on the workpiece in a global Cartesian coordinate system. For this purpose (and for computing the gravity potential in Eqn. 2a), a transformation that transforms the internal representation $q(s)$ into the Cartesian coordinates $x(s)$ for each point on the DLO is necessary.

Generally, $x(s)$ is given by

$$x(s) := \int_0^s t(\tilde{s}) \, d\tilde{s} \tag{15},$$

with as the unit tangent vector of the DLO at $\tilde{s}$ with respect to the global Cartesian system.[11]

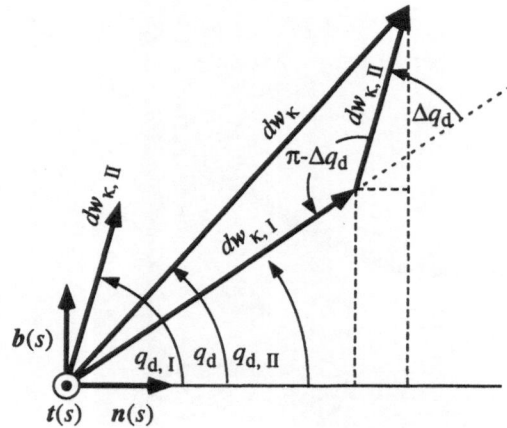

Figure 3: Adding of rotation vectors $dw_{\kappa\,\mathrm{I}}$ and $dw_{\kappa,\,\mathrm{II}}$, both lying in the b-n-plane of the accompanying trihedron at DLO position s

According to Section 3.3.1, the accompanying trihedron at position $s{+}ds$ on the DLO is derived from the accompanying trihedron at position s by performing two rotations, with the first one representing bending and the second one representing twisting. For computing $_{\mathrm{Global}}t(s)$, it is helpful to draw up the transformation matrix $_sT_{s+ds}$, with

$$_st(s+ds) =_s T_{s+ds} \cdot_{s+ds} t(s+ds),$$

i.e., to express the orientation of the tangent vector at $s{+}ds$ with respect to the accompanying trihedron at s.[12]

For this purpose, the rotation representing the bending of the DLO is further divided into three rotations:

1. First, the accompanying trihedron $\{t, n, b\}$ at s is rotated by q_d about t, aligning n with the axis of rotation in Figure 2 (top). The resulting trihedron is $\{t^{\mathrm{I}} \equiv t, n^{\mathrm{I}}, b^{\mathrm{I}}\}$.

2. Second, the trihedron obtained in step 1 is rotated by dw_κ about n^{I}. The resulting trihedron is $\{t^{\mathrm{II}}, n^{\mathrm{II}} \equiv n^{\mathrm{I}}, b^{\mathrm{II}}\}$.

3. Finally, the trihedron obtained in step 2 is rotated by $-q_d$ about t^{II}. The resulting trihedron is $\{t' \equiv t^{\mathrm{II}}, n', b'\}$, as shown in Figure 2 (bottom).

By additionally taking torsion (rotation around t') into consideration, we finally find

---

[11] In the following, a leading subscript to a vector gives the coordinate system in which the vector is expressed.

[12] Please note that $_st(s) \equiv [1, 0, 0]^{\mathrm{T}}$.

$$_sT_{s+ds} = R(t,q_d)\,R(n^I,dw_\kappa)\,R(t^{II},-q_d)\,R(t',dw_\tau) \qquad (16).$$

$R(a, \alpha)$ is the rotation matrix which transforms $_2x$ into $_1x$ with coordinate system (2) being rotated with respect to system (1) by α about axis a. In the case of Eqn. 16, all rotations are performed about elementary vectors of the trihedra. Thus, the single rotation matrices R are very simple.

Based on Eqn. 16, the transformation $_{Global}T_s$ of the accompanying polyhedron at s into global coordinates is defined recursively by

$$T(s) = {}_{Global}T_s = T(s-ds)\,_{s-ds}T_s \qquad (17).$$

Together with Eqn. 15, the global coordinates of point s on the DLO are finally given by

$$x(s) = \int_0^s T(\tilde{s}) \begin{pmatrix} 1 \\ 0 \\ 0 \end{pmatrix} d\tilde{s}\;.$$

Please note that Eqn. 17 is meaningful only for $s > 0$. Therefore, we define $T(0)$, i.e., the DLO point held by the gripper, as the transformation matrix which transforms the accompanying trihedron of the gripper into global coordinates, expressed, e.g., by Eulerian angles.

3.4 Series Expansion

Since the number $N = 3N_c$ of series coefficients that must be determined by the optimization algorithm has major influence on the computation time, the series expansion for the $q_i(s)$ should be a good approximation with few series terms. The more severely the workpiece is being deformed from its stress-free shape, the more complicated becomes this problem. Since the workpiece may generally take an arbitrary shape, it is not possible to find a series expansion that meets all possible situations. In [2, 3], Fourier series are proposed. As an alternative, we investigate the usage of Chebyshev polynomials which are often found to be a good choice for approximating unknown function analytically [7]. The polynomial of order j has the form[13]

$$Q_j(x) = \cos(j\arccos(x))\,,$$

and the series expansion for the q_j is then given by Eqn. 4. In order to meet the definition range of the arccos-function, the curve length $s \in [0 \ldots L]$ must be normalized by $x = 2s/L - 1$. If Fourier series are used, a similar normalization to the range $[-\pi \ldots \pi]$ is required.

3.5 Optimization Algorithm

In order to solve the minimization problem in determining the set c of coefficients for the series expansion, any nonlinear optimization algorithm in Multidimensions can be used. We implemented two algorithms of different complexity.

---

[13] Alternatively, the terms Q_j can be expressed by recursively defined polynomials.

The first one is the downhill simplex (DS) algorithm invented by Nelder and Mead. Here, a simplex is the geometric figure consisting of $N + 1$ points (vertices) in N dimensions including their interconnecting line segments, faces, etc. For example, in two dimensions a simplex is formed by a triangle.

The basic principle of the DS algorithm for minimizing the function $U(c)$ is seen as follows: At the beginning, an initial simplex consisting of the $N+1$ coefficient vectors $c_0, c_1, ...c_N$, is chosen arbitrarily. Then, the minimum of U is computed as follows: In each iteration, the vector c_{worst} with the highest function value U is modified according to diverse rules. In this process, the simplex is stepwise contracted and moves downhill towards the minimum of U.

Even though the downhill simplex algorithm is not very efficient, it has the following two advantages: First, it only evaluates the function U itself and does not need its (partial) derivatives. Second, it is easy to implement and is generally a good choice if the aim is "to get something working quickly" [7].

The second algorithm is the Davidon-Fletcher-Powell (DFP) algorithm as a standard variable metric method. Comparing to the DS algorithm, variable metric methods are more powerful. However, they require the evaluation not only of U, but also the vector of its first partial derivatives, ∇U, and the inverse matrix of its second partial derivatives, i.e., the inverse Hessian matrix $H^{-1} = [\nabla U]^{-1}$. Especially the computation of H^{-1} can be very time consuming. Therefore the DFP algorithm computes H^{-1} in each optimization step approximately from H^{-1} computed in the last step.

Both the DS and the DFP algorithms are standard algorithms which are described and discussed in more detail, e.g., by Press et al. [7].

4. Efficient Simulation

In this section, we investigate the computation time and the accuracy of the simulation with dependence on optimization algorithm, series expansion, and the main parameters workpiece discretization, N_L, and number of series terms, N_c. Based on this investigation, it is possible to do the simulation of DLOs efficiently, i.e., to compute the shape of the workpiece with sufficient precision in a (rather) short time. Since the simulation is based on the same principle approach as the simulation software described in [2, 3], we assume that most of the results presented here hold true for these works, too.

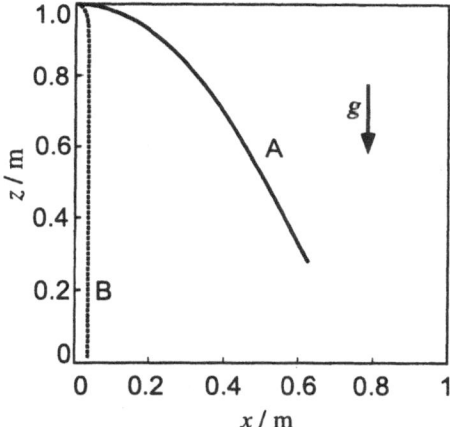

Figure 4: Two benchmarks A and B for investigating the computation precision

4.1 Computation Precision

In investigating the influence of the series expansion, the two benchmark problems shown in Figure 4 are used:

First, a copper wire of length $L = 1$ m and diameter $d = 1$ mm is fixed at one end in a Cartesian world coordinate system at $x = 0$ m, $z = 1$ m with horizontal orientation and bends due to gravity (benchmark A). Second, an additional load of 1 kg is mounted at the free end to increase the degree of bending (benchmark B). [14]

Figure 5 shows the maximum error Δx_{max} of the computed shape as a function of the number of series terms for Chebyshev polynomials and Fourier series, respectively. As reference, a computation with $N_{c, ref} = 32$ series terms is used. The number of elements for the discretization of the wire length is $N_L = 64$.

As expected, the computed shape converges with the reference shape for $N_c \to N_{c, ref}$. The deviation increases with the degree of bending of the wire. However, the number of series terms required for obtaining high accuracy is considerably lower for Chebychev terms. In this case, the maximum accuracy (given by the computational accuracy, dashed horizontal line in the figures) is obtained for $N_c \approx 10$ even for sharp bendings. If Fourier series are used, the number of required coefficients is considerably higher. Thus, we suppose that Chebyshev polynomials converge faster for typical cases.

---

[14] In reality, the deformation is mainly elastic for benchmark A while it is plastic for benchmark B. However, we consider the deformation to be purely elastic in both cases in order to have equal conditions. The plastic deformation is not relevant for the question considered here.

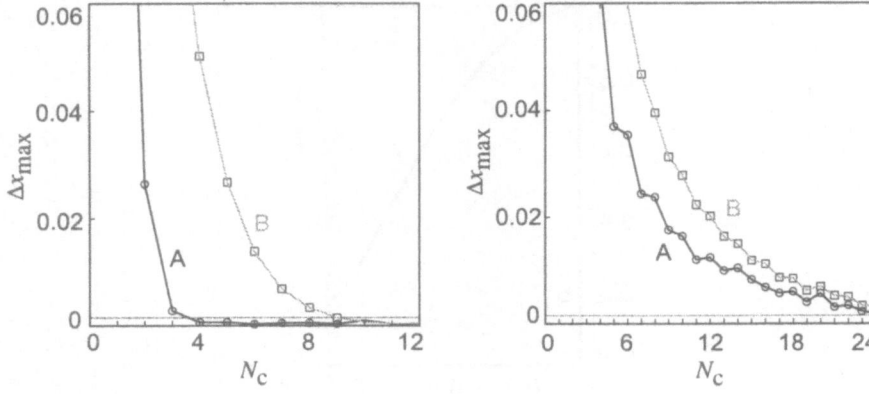

Figure 5: Maximal deviation Δx_{max} between computed and reference workpiece shape as a function of the number N_c of series terms for Chebyshev polynomials (left) and Fourier series (right)

Besides the number of series terms, the precision also depends on the discretization Δs (given by object length L and number of curve elements N_L) of the object in computing the energy integral given in Eqn. 1. For benchmark A described above, the maximal error is shown as a function of N_L in Figure 6 (for benchmark B, a similar result is obtained). The reference shape is computed with $N_{L, ref} = 960$. In this example, Chebychev series with $N_c = 16$ terms are used for approximating the q_i. With $N_L = 15$, the maximal accuracy is obtained. A further increase of N_L does not improve the accuracy.[15]

Please note that we assign one node to every discrete workpiece element. The maximum deviation Δx_{max} considered here is the maximum deviation between the computed node positions and the node positions of the reference. For the numerical integration, we approximate the object between the nodes by circular arcs.[16] Between the nodes, the difference to the reference shape may be higher than shown in Figure 5 and Figure 6.

For both optimization algorithms, Figure 8 shows the computation time as a function of the number of curve elements N_L with the number of series terms $N_c = 8$. In both cases, the computation time increases approximately linearly with N_L. Comparing with the downhill simplex algorithm, the DFP algorithm requires more evaluations of the energy integral in each iteration. Therefore, even a small differ-

---

[15] The software stores all numbers as 64 Bit floating points (standard doubles for PCs).

[16] For curved objects, an approximation by circular arcs (i.e., elements of constant curvature) are better suited than linear segments (of curvature 0). The kind of approximation especially influences the center of gravity of each segment, and, thus, the potential due to gravity (Eqn. 2a).

ence in the number of iterations has a significant impact on the total computation time. Thus, the computation time as a function of N_L is less smooth for the DFP algorithm than for the downhill simplex algorithm.[17]

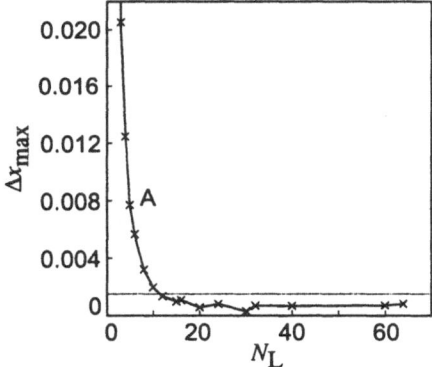

Figure 6: Maximal deviation Δx_{max} between computed workpiece shape and reference as a function of the number N_L of curve elements for benchmark A

4.2 Computation Time

The effort required for computing the shape of the workpiece is mainly determined by the combination of the following factors: Number of curve elements, N_L, number of series terms, N_c, and optimization algorithm for computing the energy minimum according to Eqn. 1.

For investigating the computational effort, benchmark C shown in Figure 7 is used: The copper wire described above is gripped at one end point with gripper position $x = 0$ m, $z = 1$ m. With the gripper orientation being initially horizontal, the gripper is rotated by 180° about the y-axis with a stepsize of 10° and back to the initial position. In each experiment, the total time for simulating the 36 object positions is measured. The computation is performed on a 133 MHz Pentium PC with 64 Mbytes RAM using LINUX as operating system.

Figure 9 shows the measured computation time as a function of the number of series terms N_c with object discretization $N_L = 32$. Obviously, gradient methods as the DFP algorithm are especially powerful if the number of coefficients to be determined is large. However, in the previous section it is shown that approximately 10 series terms are generally sufficient if $q(s)$ is approximated by Chebychev polynomials. Therefore, the downhill simplex algorithm is not only easier to implement but also faster in typical cases.

---

[17] Please note that the computation time is not generally smaller for the downhill simplex algorithm. This also depends on the number N_c of series terms.

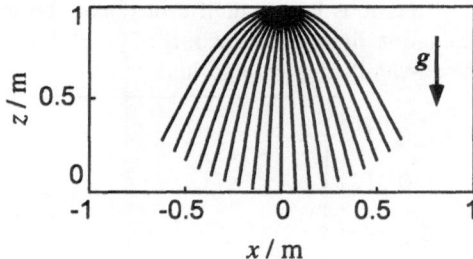

Figure 7: Benchmark C for investigating the computation time

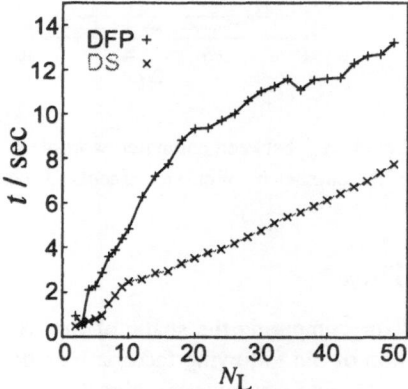

Figure 8: Computation time as a function of the number of curve elements N_L for downhill simplex (DS) and Davidon-Fletcher-Powell (DFP) algorithm

4.3 Parallel Computation

The previous section shows that a short computation time can be obtained by an appropriate selection of computation parameters and optimization algorithm. If a further reduction of the computation time is required, e.g., for real-time computation in combination with sensor evaluation, parallel computation can be considered.

In this context, we need to distinguish two different situations, which are discussed in the following.

1. Different shapes of the workpiece are independent from each other (*independent computation*).

2. The shape computed in each step depends on the shape computed in the previous step (*dependent computation*).

We implemented a parallel version of the simulation software on a workstation cluster, consisting of 9 PCs, each with 133 MHz Intel Pentium processors and 128

Mbytes memory. The parallel communication is established by an Ethernet based bus network (see [8] for details).

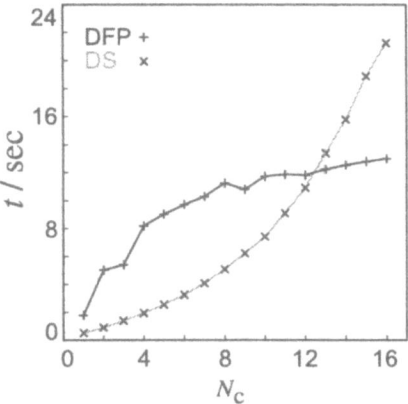

Figure 9: Computation time as a function of the number of series terms N_c for downhill simplex (DS) and Davidon-Fletcher-Powell (DFP) algorithm

4.3.4 Independent Computation

For independent computation, at least the following two basic conditions must be met: The simulation is performed (quasi-)static and there is no plastic deformation. Under these circumstances, the workpiece shape can be computed in parallel for different positions of the gripper trajectory (starting the optimization algorithm always with the same initial guess for $q(s)$). However, it is more favorable to use the result of a previous step as initial guess since the discrepancy to the actual shape is typically smaller in this case. As expected, the resulting speedup is almost linear, as shown in Figure 10 for the benchmark C given in Figure 7.

However, if any interactions between workpiece and obstacles have to be considered, the computed object shapes may not be valid, even if the conditions given above are met. This problem is demonstrated in Figure 11 (left), with the following benchmark D. The object is moved downwards (into direction MD, direction of gravity) and collides with an obstacle. All steps in the simulation are computed independently from each other, the minimization algorithm is always started with an undeformed workpiece as initial guess. For the first four steps, the object shape is correctly simulated, but in the fifth step the object "jumps" to the lower side of the obstacle, which is obviously incorrect. This is caused by the fact that the optimization algorithm seeks for the minimum energy which is next to the initial guess. Starting with an undeformed object, the algorithm always finds the global minimum in this example.

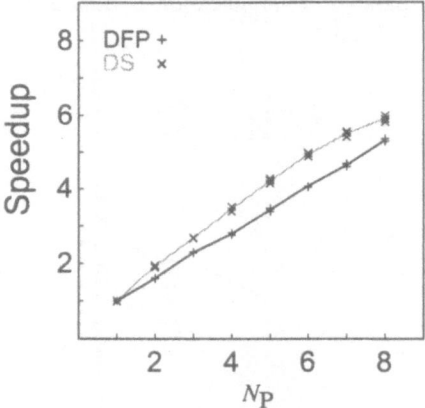

Figure 10: Measured speedup for independent computations using downhill simplex (DS) and Davidon-Fletcher-Powell (DFP) algorithm with N_p being the number of processors (each experiment performed three times)

The correct result is obtained if the object shape computed in step $i-1$ is taken as initial guess in step i, as shown in Figure 11 (right). In this case, the algorithm finds the local minimum which is next to the shape computed in the previous step. The parallel computation of several points of the trajectory is obviously not possible in this case.

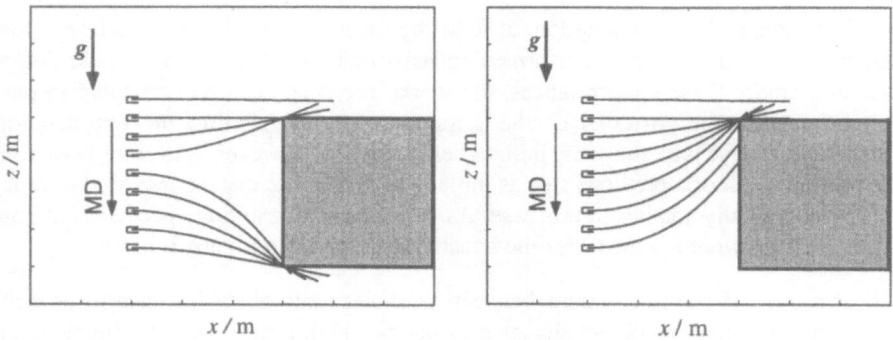

Figure 11: Benchmark D: Obstacle interaction of a deformable linear object while being moved into direction MD for independent computation of all steps (Steps 1 to 9 upwards down). Left: false result for step 5 to 9. Right: correct result

It is not always necessary to use the result of step $i-1$ as initial guess in step i in those cases. It is also possible to use a shape computed a few steps in the past. Thus, correct results can be achieved without completely losing the advantage of parallel computation. However, the number of computations to be computed in parallel must be chosen carefully if interaction with obstacles is likely to occur.

4.3.5 Dependent Computation

If the object is deformed plastically while being handled by the robot, or if the simulation needs to be performed dynamically, the simulation of step i requires the results obtained in step $i-1$ (for example the velocity of each mass element) as input data for the computation. Therefore, the approach given above is not feasible.

Generally, it may be assumed that the final solution Q_i in simulation step i (with $t = i \Delta t$) can be computed fast, if the difference between the actual minimal energy and the initial guess used for the numerical optimization is small.[18] Based on this idea, we can compute a good initial guess for the steps $i+1$, $i+2$, ..., while computing the correct result Q_i for step i.

Let us assume the case of performing a dynamic simulation and having two independent computation tasks T_1 and T_2. Given a known initial value Q_0 (i.e., the position and velocity of each workpiece point) of the object at time $t = 0$, we use the following algorithm:

Based on Q_0, T_1 and T_2 compute simultaneously two shapes $Q_{1, \Delta t}$ and $Q_{2, 2\Delta t}$ in simulation step $i = 1$. $Q_{1, \Delta t}$ is computed with timestep Δt, $Q_{2\ 2\Delta t}$ is computed with timestep $2\Delta t$. While $Q_{1, \Delta t}$ is the correct solution for $t = \Delta t$, i.e., $Q_{1, \Delta t} = Q_1$, $Q_{2\ 2\Delta t}$ is an approximation of Q_2 at (shape of the workpiece at $t = 2\Delta t$). It is only an approximation, since the correct computation of Q_2 requires Q_1 as input data.

In simulation step $i = 2$, task T_2 continues its computation for Q_2, using $Q_{2\ 2\Delta t}$ as inital guess and Q_1 as input data. The result, $Q_{2, \Delta t} = Q_2$, is the correct solution for $t = 2\Delta t$. Simultaneously, T_1 computes an approximation $Q_{3, 2\Delta t}$ for Q_3, and so on.

This approach can be easily extended to an arbitrary number of independent computation tasks. While one task computes the correct solution for $t = i \Delta t$ (using the timestep Δt), the other tasks compute approximations for $t = \{(i+1)\Delta t, (i+2)\Delta t, ...\}$.

However, the assumption that a good initial guess results in a short computation time for the energy minimum is not always true, but depends on the optimization algorithm. On the one hand, the downhill simplex algorithm requires $3N_c + 1$ independent guesses for each of the $3N_c$ parameters. Having just one good (maybe almost optimal) guess from the previous steps does not significantly simplify the problem. Accordingly, the possible speedup is rather low.

Gradient-based algorithms, such as DFP, on the other hand, come close to the minimum rather fast even if the initial guess is bad, but it requires many iterations to finally determine the minimum with the desired accuracy. Therefore, the influence of a good initial guess is rather small, resulting in a low speedup. However, if the guess computed in step $i-1$ is the actual shape of the workpiece, the DFP algorithm terminates immediately. This is the case if there is no plastic deformation in a step or if the acceleration of all mass elements is constant in time, respectively. Here, the speedup is lower than for independent computation, but also linear.

---

[18] $Q_i(s)$ consists of the functions representing the workpiece shape $q_1(s)$, $q_2(s)$, $q_3(s)$, and the (Cartesian) velocity $v(s)$ of each workpiece point in step i.

Figure 12 shows the measured speedup for a dynamic computation of the benchmark C as shown in Figure 7 for both minimization algorithms. Due to the characteristics of the downhill simplex algorithm described above, the speedup is almost negligible. For the DFP algorithm, a maximum speedup of about two is obtained for three computation tasks. If the number of tasks is further increased, the speedup decreases due to the following reasons: First, the more computation tasks we use, the more guesses we compute for future simulation steps. However, if a guess for step $i+k$ is computed in step i, the significance of the guess decreases with increasing k. Therefore, the additional benefit of the tasks becomes smaller from task to task. Second, the effort required for communication increases with the number of computation tasks.

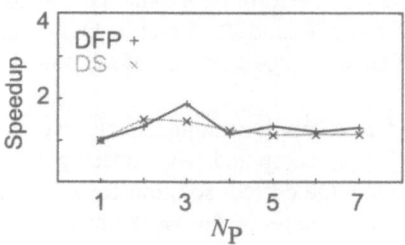

Figure 12: Measured speedup for dynamic (dependent) computation using downhill simplex (DS) and Davidon-Fletcher-Powell (DFP) algorithm with N_p being the number of processors (each experiment performed three times)

Comparing with the resulting speedup, it is found that the effort for the parallel computation is too high in the case of dependent computations. Some additional possibilities for parallelizing sub-tasks, e.g., the computation of the energy integral, have been considered, but have not been implemented because we expected speedup to be poor.

5. Inverse Simulation

5.1 Approach

So far, we have discussed the problem of simulating the behavior of the work-piece if the gripper trajectory (and possibly other boundary conditions like obstacles) are given. We call such problems *"Direct Simulation Problems"*. They occur, e.g., when different handling strategies are compared or when the impact of the material parameters is studied.

However, if we think of using a simulation system for robot programming (i.e., generating the gripper trajectory for a given task), the problem is just inverse: Given some boundary conditions concerning the position and shape of the workpiece, the gripper trajectory shall be computed. We call this kind of problems "*Inverse Simulation Problems*". Till now, it has not been investigated systematically.

The simplest example is the threading of a DLO through a cut-out, e.g., in a sheet metal. The optimal solution for this task is a gripper trajectory which meets the following conditions for the complete threading process, guaranteeing maximal tolerance to all kinds of uncertainties or distortions:

- The DLO pierces the sheet metal plane at the center point P_{Goal} of the cut-out, and

- the orientation of the DLO at P_{Goal} is aligned with the normal n_{Goal} of the sheet metal.

To solve this problem, we recall the solution to the direct simulation problem, as described above. To do this, we introduce a penalty function $U_{Penalty}$, describing the deviation between given boundary conditions and actual DLO shape. By changing the position $P_{Gripper}$ and orientation $n_{Gripper}$ of the gripper, $U_{Penalty}$ is iteratively minimized. This approach leads to the following algorithm.

1    $P_{Gripper} := P_{Gripper,\,0};$ {Initial guess for gripper position & orientation}
     $n_{Gripper} := n_{Gripper,\,0};$
2   **repeat** {Main loop for inverse simulation problem}
3   *Solved* := **false**;
4   *Deviation* := $U_{Penalty}(Q, P_{Goal}, n_{Goal})$;
5   **if** *Deviation* = 0 **then** {solved}
        Solved := **true**
    else {not solved}
        $P_{Gripper} := P_{Gripper} + \Delta P_{Gripper};$ {varying gripper position
        $n_{Gripper} := n_{Gripper} + \Delta n_{Gripper};$ and orientation}
6   **until** *Solved*;

In this algorithm, Q is the shape of the workpiece, given by the vectors of all line elements, $\Delta P_{Gripper}$ and $\Delta n_{Gripper}$ are the change of the gripper position and orientation in each iteration.

In this approach, we use two interlaced optimization processes. The "inner" one for computing the DLO shape for a given gripper position (direct simulation problem), and the "outer" one for determining the gripper position which solves the inverse problem. With this approach, it is possible to consider any kind of boundary conditions by an appropriate selection of $U_{Penalty}$.

5.2 Solution for Important Special Cases

However, there are important special cases in which the problem can be solved much easier. As long as no interaction between workpiece and environment has to be considered, the shape of the DLO is only affected by the orientation of the gripper with respect to gravity, but it is independent of the absolute gripper position.

Therefore, the following strategy can be used for solving the inverse simulation problem:

Given P_{Goal}, n_{Goal} and the line element $n \in [0, 1, ..., N_L-1]$ of the DLO which shall pierce the cutout, $P_{Gripper}$ is fixed at an arbitrary position, e.g., the origin of the global coordinate system, and $n_{Gripper}$ is varied until the DLO tangent is aligned with n_{Goal}. With P_n' being the position of DLO element n, the gripper must then be displaced to

$$P_{Gripper} = P_{Goal} - P_n'$$

in order to finally solve the problem. With this approach, the algorithm given above is applied only to determine the gripper orientation, while the gripper position can be computed directly. As long as the curve describing the shape of the DLO lies within a plane (which holds approximately true in many cases), the problem does not have to be regarded in three dimensions, but is two-dimensional. In this case, only one angle θ is required for determining the gripper orientation. Thus, the "outer" optimization problem in the algorithm given above is an one-dimensional optimization for θ. Being t_n the tangent vector of the DLO in point n, a quadratic penalty function

$$U_{Penalty} = c \, (\arccos(t_n n_{Goal}))^2$$

is found to be suited for the optimization, with c being an adjustment constant.

The left column of Figure 13 shows simulation examples for the threading of flexible beams with different bending rigidity through a cut-out with the normal of the sheet metal plane being perpendicular to gravity. The increase of bending due to gravity for decreasing bending rigidity (from top to bottom) is obvious.

Please note that in Zheng et al. [1], the task of inserting a flexible beam into a bore hole of approximately equal diameter is regarded. In that case, the gripper trajectory is found to be equivalent to the deflection curve of the bending beam. However, this is no general solution for the inverse simulation problem. For the task described in [1], the already inserted portion of the beam does not have to be considered any longer, because its weight is compensated by the walls of the bore hole. In other tasks, e.g., the threading task considered here, the complete beam is deflected by gravity while the task is performed. The relation between the gripper trajectory for the threading task and the insertion into a bore hole is shown in the right column of Figure 13. In each figure, the upper curve represents the DLO shape for the beginning of the threading process, i.e., the free end of the DLO being at position P_{Goal}, having the orientation n_{Goal}. For the insertion task described by Zheng et al., this curve defines the complete gripper trajectory. In the contrary, the lower curves in the figures give the gripper trajectory for the threading tasks shown in the left column. Obviously, the trajectory is different for both tasks, though both $P_{Gripper}$ and $n_{Gripper}$ are identical in the beginning.

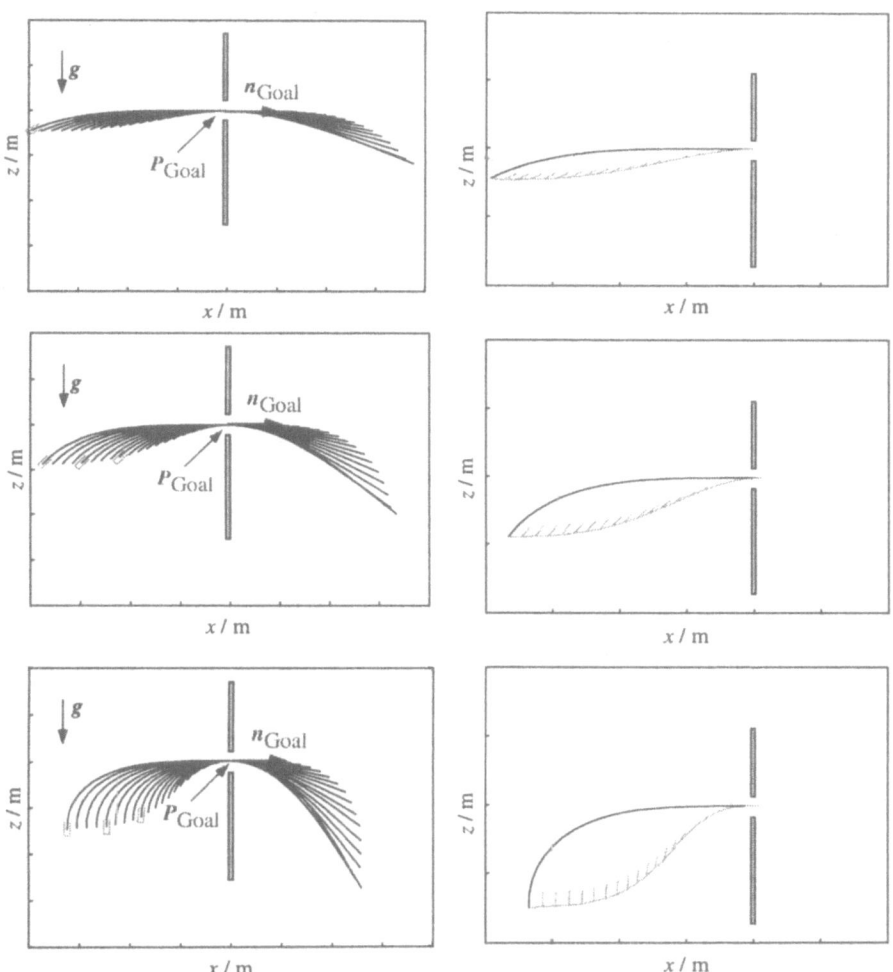

Figure 13: Threading of a bending wire of length $L = 1$ m through a cut-out with center $x_{Goal} = 0$ m, $z_{Goal} = 1$ m for different bending rigidities (decreasing from top to bottom). Figures left: Wire shape during the threading process. Figures right: Initial wire shape (upper curve) and gripper trajectory with gripper orientation indicated by the short lines (lower curve).

In the task of threading through a cut-out with a given orientation, the required position and orientation of the gripper are defined unequivocally for each point of its trajectory. Instead of giving the position and orientation of one point as constraints, it is also possible to give two goal positions P_1 and P_2 without orientation. An example for this kind of task is the threading through two cut-outs. In this case, it is desired to guide the DLO simultaneously through the centers of both cut-outs. Again, it is not necessary to determine both position and orientation by means of numerical optimization, but the gripper displacement can be computed directly. An example is shown in Figure 15.

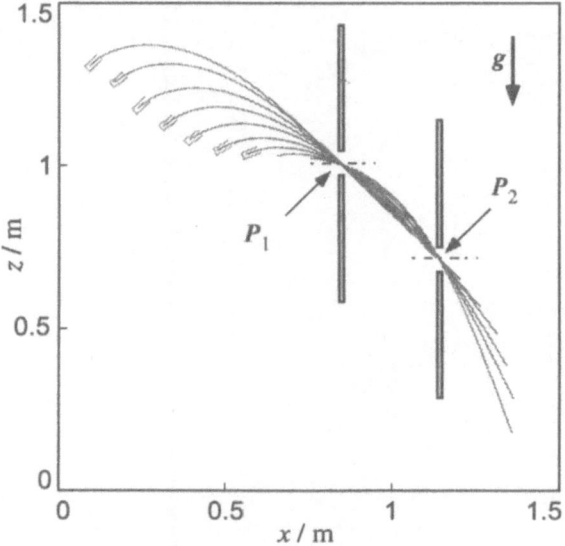

Figure 14: Example for inverse simulation with two desired goal positions P_1 and P_2:
Threading through two cut-outs

5.3 Outlook

Comparing with the general algorithm described above, the computation can be simplified in the cases discussed here because the workpiece shape is independent from the gripper position. This holds true as long as the DLO is deformed only by gravity and no direct contact between DLO and environment occurs. If this condition is not fulfilled, the situation becomes more complex.

As example, Figure 15 shows the numerical simulation of an experiment described by Henrich et al. [9]: The DLO is in contact with an obstacle and is being deformed by the contact force. The gripper is moved on a trajectory which iteratively reduces the DLO length between gripper and contact point, while the DLO orientation n_{Goal} is kept constant.[19]

However, another experiment described in [9] using the same setup states that for an arbitrary contact point along the DLO length and a given DLO orientation in the contact point, a gripper trajectory exists which

- retains the contact during the motion, and

---

[19] In this example, the gripper trajectory is given by a straight line, connecting the initial gripper position and the contact point. The gripper orientation is constant [9].

- ensures that neither the contact point along the DLO length nor the orientation of the DLO in the contact point is altered.

This means that for any given contact point along the DLO and given orientation at the contact point, the solution of the inverse problem is not unambiguous, but the number of solutions is infinite. This is demonstrated in the example shown in Figure 16 for a horizontal orientation n_{Goal} of the DLO in the contact point.

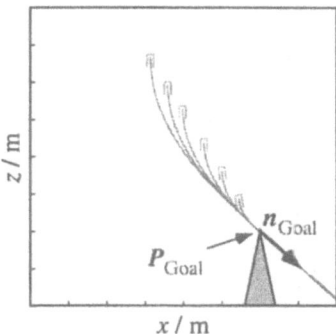

Figure 15: Example for inverse modeling problem if contact between workpiece and environment must be considered. The DLO length between gripper and contact point P_{Goal} is iteratively reduced while the orientation in P_{Goal} is kept constant.

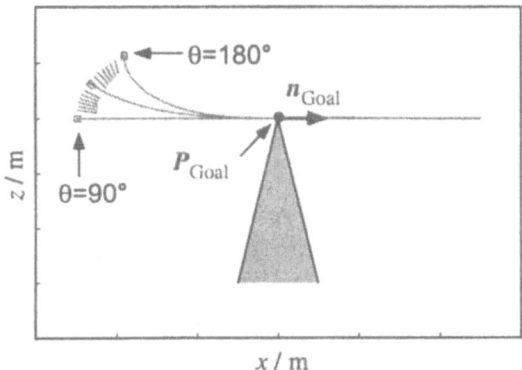

Figure 16: Different solutions for the inverse modeling problem with contact between workpiece and obstacle. All of the gripper positions shown here (and all points on the trajectory connecting them) are valid solutions for the inverse simulation problem.

Because it is generally not clear what solution the algorithm given above will find, it is required to add additional boundary conditions in such cases, which guarantee an unequivocal solution. The problem of solving the inverse simulation problem with contact between workpiece and environment is currently being investigated.

6. References

[1]   Zheng, Y. F., Pei, R., Chen, C., "Strategies for automatic assembly of deformable objects", In: Proc. 1991 Int. Conf. on Robotics and Automation, vol. 3, pp. 2598-2630, Sacramento, USA, April 1991.

[2]   Hirai, S., Wakamatsu, H., Iwata, K., "Modeling of deformable thin parts for their manipulation", In: Proc. 1994 Int. Conf. on Robotics and Automation, vol. 4, pp. 2955-2960, San Diego, USA, May 1994.

[3]   Wakamatsu, H., Hirai, S., Iwata, K., "Modeling of linear objects considering bend, twist and extensional deformations", In: Proc. 1995 Int. Conf. on Robotics and Automation, vol. 1, pp. 433438, Nagoya, Japan, May 1995.

[4]   Wakamatsu, H., et al., "Dynamic analysis of rodlike object deformation towards their dynamic manipulation", In: Proc. 1997 IEEE/RSJ Int. Conf. on Intelligent Robots and Systems" (IROS'97), pp. 196ff, Grenoble, France, Sept. 1997.

[5]   Ritz, W.: "Über eine neue Methode zur Lösung gewisser Variationsprobleme der mathematischen Physik". Journal für die reine und angewandte Mathematik", vol. 135, 1909.

[6]   Nakagaki, H., et al., "Study of deformation and insertion tasks of a flexible wire", In: Proc. 1997 Int. Conf. on Robotics and Automation, vol. 3, pp. 2397-2402, Albuquerque, USA, April 1997.

[7]   Press, W. H., et al.: "Numerical recipes in C". Second edition, Cambridge University Press, 1992.

[8]   Wurll, C., Henrich, D.: "Ein Workstation-Cluster für paralleles Rechnen in Robotik-Anwendungen" (A workstation cluster for parallel processing in robotic applications). In: Proceedings der 4. ITG/GI-Fachtagung Arbeitsplatz-Rechensysteme (APS'97), Universität Koblenz-Landau, Germany, 21.-22. May, 1997, pp. 187 - 196.

[9]   Henrich D., Ogasawara T., and Wörn H.: "Manipulating deformable linear objects: Contact states and point contacts". In: Proc. 1999 IEEE International Symposium on Assembly and Task Planning (ISATP'99), Porto, Portugal, July 21-24, 1999.

Chapter 3

Planning and Control Strategies

Section 3.1

Indirect Simultaneous Positioning of Extensible Deformable Objects

T. Wada

Abstract.   Positioning operations of multiple points on an extensible deformable object will be discussed. In many operations that deal with extensible deformable objects such as rubber parts and textile fabrics, multiple points on the object should be guided to their desired locations. In addition, we cannot operate the guided points directly because of collision among positioning devices and other devices. Thus, the operation should be performed by controlling other points on the object. This operation is referred to as indirect simultaneous positioning of multiple points on an extensible deformable object. In this article, control problems for indirect simultaneous positioning are treated.

In order to perform indirect simultaneous positioning by a mechanical system, an object model is indispensable. However, it is difficult to build an exact model of a deformable object due to strong nonlinearity including friction, hysteresis, and parameter variation. To overcome this problem, I will propose a robust control strategy based on a coarse model of deformable objects. I will build a coarse model of an object for its positioning and will develop a control method robust to the discrepancy between the object and its model.

First, a coarse model of extensible deformable objects is developed for their positioning. Second, indirect positioning is formulated. Then, the condition that a given positioning can be realized is derived. A novel control method for the indirect positioning with a visual sensor is proposed. Experimental results and theoretical analysis will show the robustness of the proposed method against the discrepancy between an object and its coarse model.

1 Introduction

Many manipulative tasks deal with deformable objects such as textile fabrics, rubber parts, paper sheets, and food products. Most these operations strongly depend on skilled human workers. In general, it is difficult to achieve such operations by mechanical systems such as robot manipulators due to complexity of their deformation characteristics such as nonlinear elasticity, friction, hysteresis, and parameter variations.

Some researches on manipulations of deformable objects have been conducted. Especially for automated manufacturing of textile fabrics, many researches have been done [1]. For such purposes, Hall et al.[2], Kemp et al.[3], and Ono et al.[4] have dealt with methods to separate a fabric piece from a stack. Some projects have developed automated sewing systems [5] [6] [7]. Ono et al. [8] also have derived a strategy for unfolding a fabric piece based on cooperative sensing of touch and vision. In these researches, since their approaches are for a specific task, thus it is difficult to apply the results to other different tasks with a systematic manner. Also, deformation of fabrics is not controlled actively while some of them deal with their position and orientation. On the other hand, some researches have tried to deal with on more general deformable object with systematic manners as follows. Hirai et al. [9] have proposed a method for modeling linear objects based on their potential energy and analyzed their static deformation. Wakamatsu et al. [10] have analyzed grasping of deformable objects and introduced bounded force closure. Their approach is static, control of manipulative operations is out of consideration. Howard et al. [11] have proposed a method to model elastic objects by the connections of springs and dampers. Then, a method to estimate the coefficients of the springs and dampers has been developed by recursive learning method for grasping. This study has focused on model building. Thus, control problems for manipulative operations have not been investigated. Sun et al. [12] have studied on the positioning operation of deformable objects using two manipulators. They have focused on the control of the object position while deformation control is not discussed.

In this article, I focus on the manipulations of deformable objects such as textile fabrics and rubber parts. These deformable objects. Such deformable objects represent not only bending deformations but also deformations along stretch directions. We refer to these deformable objects as extensible deformable objects. Fig.1 illustrates a special sewing operation of knitted fabrics, which is called linking, as an example of manipulations of extensible deformable objects [13]. In this operation, each corresponding pair of knitted loops (points) have to be matched precisely before sewing. In this case, we cannot operate the points to be positioned due to collision between positioning mechanisms and a sewing needle. Fig.2 depicts essences of the linking operations in a general form. Namely, in this operation, multiple points on an object should be guided to their desired location simultaneously as illustrated in Fig.2. The positioned points cannot be operated directly. Therefore, we have to realize the operation by controlling other points on the deformable objects. We call such operation "indirect simultaneous positioning of multiple points on an extensible deformable object" [14]. In this article, I deal with the indirect simultaneous positioning.

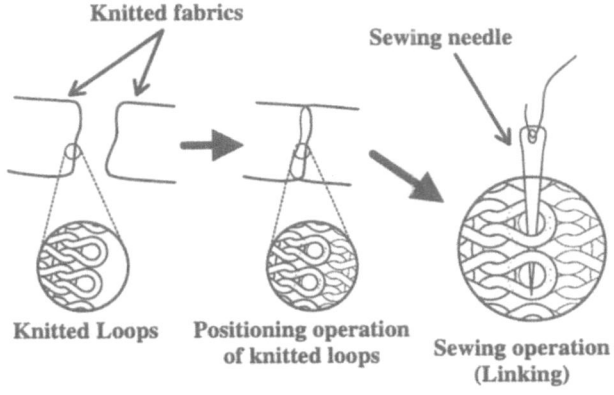

Figure 1: Sewing operation of knitted fabrics

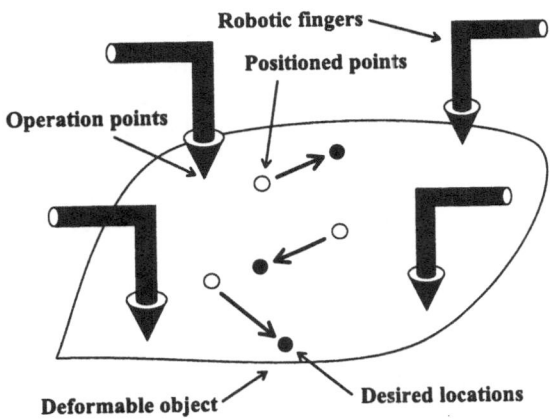

Figure 2: Indirect positioning of deformable object

In order to realize indirect simultaneous positioning, object model is indispensable. However, it is difficult to build exact model of the deformable objects in general due to nonlinear elasticity, friction, hysteresis, parameter variations, and other uncertainties. These are the main difficulties in manipulating deformable objects. To solve this dilemma, I propose to utilize a coarse object model to derive task strategies. In our approach, I first build a coarse model of a manipulated object. Then, the task is analyzed based on the proposed coarse model. One of the advantages using the coarse object model is that we can analyze the task and may realize its essence relatively easily. Based on the results of the analysis, we can derive a control law and task planning method that are robust to discrepancy between the object and its coarse model.

In this article, I will firstly propose a coarse model of extensible deformable objects. Next, indirect positioning will be analyzed based on the coarse model. As the result, I will derive conditions to examine whether the given positioning is feasible or not. Then, I will propose a control method robust to model errors based on the coarse model. Experimental results show the validity of the proposed method and the effects of the parameter errors on the convergence.

2 Formulation Of Indirect Positioning

2.1 Modeling Of Extensible Deformable Objects

First of all, the model of deformable objects is proposed in order to formulate problems. In general, force–displacement characteristics of the deformable object are nonlinear. In addition, there exist hysteresis properties in some materials. However, in this article, I model the static deformation characteristics of the object as connections of linear springs as shown in Fig.3 in order to develop a simple model. But, note that this simple model can describe a class of nonlinear characteristics because directions of springs change with object deformation even though each element consists of linear spring.

For simplicity, I deal with two dimensional deformable objects such as textile fabrics. Suppose that the deformable objects are dealt with on a smooth table. Thus, we can neglect the gravity effect exerted on the objects and friction forces between the objects and table. In the proposed model, each mesh point is connected by vertical, horizontal, and diagonal springs as shown in Fig.3. In the model, we assume that the object deforms in a two-dimensional plane. Position vector of the (i, j)-th mesh point $p_{i,j} = [x_{i,j}, y_{i,j}]^T$ $(i = 0, \cdots, M; j = 0, \cdots, N)$ is utilized in order to describe translations, rotations, and deformations of the object. Let l_x, l_y, and l_θ be natural length of horizontal, vertical, and diagonal springs, respectively. Coefficients k_x, k_y, k_θ are spring constants of horizontal, vertical, and diagonal springs. Assume that no moment is exerted on mesh points. Then, the resultant force exerted on mesh point $p_{i,j}$ can be described as eq.(1).

$$F_{i,j} = \sum_{k=1}^{8} F_{i,j}^k = -\frac{\partial U}{\partial p_{i,j}} \tag{1}$$

U denotes whole potential energy of the object. Then, function U can be calculated by sum of all energies of springs [14]. Here, we assume that the shape of the object is dominated by eq.(1). Then, we can calculate the deformation of the object by solving eq.(1) under given constraints. Note that the following discussions are valid even if the object has an arbitrary three-dimensional shape by modeling the object similarly. Details have been reported in [14].

The dynamic behavior of the deformable objects can be described if we give a mass to each mesh point and dampers are installed between the mesh points

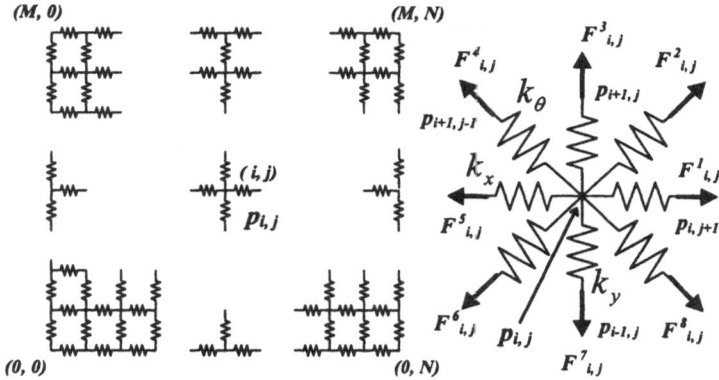

Figure 3: Spring model of deformable object

in parallel to the springs. In this article, such masses and dampers are omitted because I utilize static characteristics of the object in deriving control strategy.

The proposed model cannot describe hysteresis property that is seen in practical deformations of the the objects, for instance textile fabrics. However, it has been shown that the model can represent the fundamental deformation characteristics through experiments [15]. In this research, I regard that the proposed model is suitable to the positioning tasks. Then, I derive a method to realize the positioning tasks based on the model.

2.2 Problem Description

Here, I classify mesh points $p_{i,j}$ into the following three categories(see Fig.4) in order to formulate indirect simultaneous positioning.

operation points: are defined as the points that can be manipulated directly by robotic fingers. ($\triangle$)

positioned points: are defined as the points that should be positioned indirectly by manipulating operation points appropriately. ($\bigcirc$)

non-target points: are defined as the all points except the above two points. (others in Fig.4)

Let the number of operation points and of positioned points be m and p, respectively. The number of non-target points is $n = (M + 1) \times (N + 1) - m - p$. Then, r_m is defined as a vector that consists of coordinate values of the operation points. Vectors r_p and r_n are also defined for positioned and non-target points in the similar way. Eq.(1) can be rewritten as eqs.(2),(3) using r_m, r_p, and r_n.

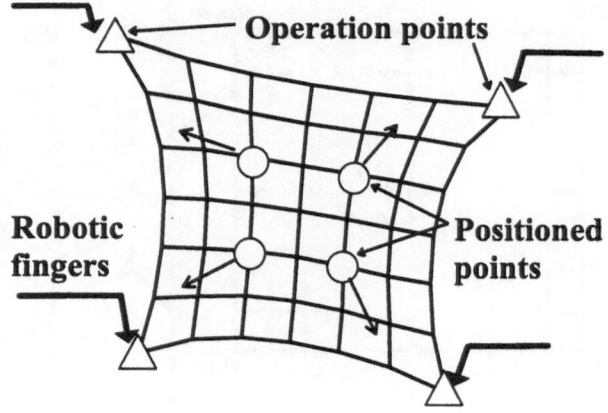

Figure 4: Classification of mesh point

$$\frac{\partial U(\boldsymbol{r}_m, \boldsymbol{r}_n, \boldsymbol{r}_p)}{\partial \boldsymbol{r}_m} - \boldsymbol{\lambda} = 0, \tag{2}$$

$$\left[\begin{array}{c} \frac{\partial U(\boldsymbol{r}_m, \boldsymbol{r}_n, \boldsymbol{r}_p)}{\partial \boldsymbol{r}_p} \\ \frac{\partial U(\boldsymbol{r}_m, \boldsymbol{r}_n, \boldsymbol{r}_p)}{\partial \boldsymbol{r}_n} \end{array} \right] = 0 \tag{3}$$

where a vector $\boldsymbol{\lambda}$ denotes a set of forces exerted on the object at the operation points $\boldsymbol{r}_m$ by robotic fingers.

Note that the external forces $\boldsymbol{\lambda}$ can appear only in eq.(2), not in eq.(3). This implies that no external forces are exerted on positioned points and non-target points. These equations represent characteristics of indirect simultaneous positioning of deformable objects.

Let us consider the following task:

[Task]   *Assume that the configuration of robotic fingers and the positioned points on an object are given in advance. Then, the positioned points $\boldsymbol{r}_p$ are guided to their desired location $\boldsymbol{r}_p^d$ by controlling operation points $\boldsymbol{r}_m$ appropriately.*

In order to realize the given task, an object model is indispensable since we have to predict directions of displacements of positioned points during the positioning. Then, the proposed model is useful for this purpose. However, in general, the model errors cannot be ignored in modeling deformable objects due to hysteresis, friction, parameter variations, and other uncertainties. Thus, a model inversion approach is not effective. Therefore, it is important to develop a control method that is robust to the error between the object and its model.

3 Analysis Of Indirect Positioning

3.1 Infinitesimal Relation Among Positioned Points And Operation Points

In this section, I analyze indirect simultaneous positioning based on the proposed coarse model. Let us derive infinitesimal relation among positioned points and operation points. Now, consider a neighborhood around an equilibrium point $r_0 = [r_{m0}^T, r_{p0}^T, r_{n0}^T]^T$. We can obtain the following equation by linearlizing eq.(3) around the equilibrium point.

$$A\delta r_m + B\delta r_n + C\delta r_p = 0 \quad !! \tag{4}$$

where

$$
A \triangleq \left[\begin{array}{c} \frac{\partial^2 U}{\partial r_m\, \partial r_p} \\ \frac{\partial^2 U}{\partial r_m\, \partial r_n} \end{array} \right]\Bigg|_{r0} \quad \in R^{(2p+2n)\times 2m}
$$

$$
B \triangleq \left[\begin{array}{c} \frac{\partial^2 U}{\partial r_n\, \partial r_p} \\ \frac{\partial^2 U}{\partial r_n\, \partial r_n} \end{array} \right]\Bigg|_{r0} \quad \in R^{(2p+2n)\times 2n} \tag{5}
$$

$$
C \triangleq \left[\begin{array}{c} \frac{\partial^2 U}{\partial r_p\, \partial r_p} \\ \frac{\partial^2 U}{\partial r_p\, \partial r_n} \end{array} \right]\Bigg|_{r0} \quad \in R^{(2p+2n)\times 2p}.
$$

Vector δr_m is defined as an infinitesimal deviation of the operation points from their equilibrium points. Vectors δr_n and δr_p are defined in the similar way.

3.2 Feasibility Of Indirect Positioning

At first, we can obtain the following theorems.

Theorem 1 **There exist infinitesimal displacements of operation points δr_m corresponding to arbitrary infinitesimal displacements δr_p, if and only if,**
rank$[A\ B] = 2p + 2n$ **is satisfied.**

[proof]
By transforming eq.(4), eq.(6) is obtained.

$$
F \left[\begin{array}{c} \delta r_m \\ \delta r_n \end{array} \right] = -C\delta r_p \tag{6}
$$

where $F = [A\ B]$.
Let $S(\cdot)$ be vector space that is spanned by column vectors of a matrix. In eq.(6), there exists δr_m corresponding to arbitrary δr_p, if and only if,

$$S([A\ B]) \supseteq S(C) \tag{7}$$

is satisfied according to conditions that there exist solutions of linear equations.

Now, we investigate the characteristics of matrices B and C. Eq.(4) can be rewritten as:

$$A\delta r_m = -[B \ C] \begin{bmatrix} \delta r_n \\ \delta r_p \end{bmatrix}. \tag{8}$$

According to characteristics of spring models, the infinitesimal displacements of positioned points δr_p and those of non-target points δr_n can be determined uniquely, corresponding to a given infinitesimal displacement of operation points δr_m for any number and configurations of the operation points. This yields the following.

$$\text{rank}[B \ C] = 2p + 2n \quad \forall m, \ \forall r_m. \tag{9}$$

Then, the next equation is satisfied.

$$S(B) \cap S(C) = \{0\}. \tag{10}$$

Considering eq.(10) with eq.(7) yields

$$S(A) \supseteq S(C). \tag{11}$$

With eqs.(9), (10), and (11), eq.(7) can be rewritten as $\text{rank}[A \ B] = 2p+2n$. Thus, the proof has been completed.

In addition, Theorem 1 needs the following result.

Result 1 ! *The number of the operation points must be greater than or equal to that of the positioned points in order to realize any arbitrary displacement δr_p , that is, $m \geq p$.*

In the case that the number of the operation points is equal to that of the positioned points, that is, $m = p$ is satisfied, Theorem 1 can be rewritten as follows:

Theorem 2 ! *In the case of $m = p$, there exist displacements of the operation points δr_m corresponding to any displacements of positioned points δr_p and these are determined uniquely, if and only if, $\det[A \ B] \neq 0$.*

3.3 Example

The following example illustrates the validity of the proposed conditions.

Fig.5-(a) and (b) show the indirect positioning of three positioned points by three operation points. In these examples, the deformable object is modeled by 4×4 mesh points. In the case of Fig.5-(a), $\det[A, \ B] \neq 0$ is satisfied while $\det[A, \ B] = 0$ in the case of (b). Thus, we can conclude that the positioning operation is feasible in the case of (a) while cannot achieve in the case of (b). These results are corresponding to our intuition. These examples show the validity of derived theorems.

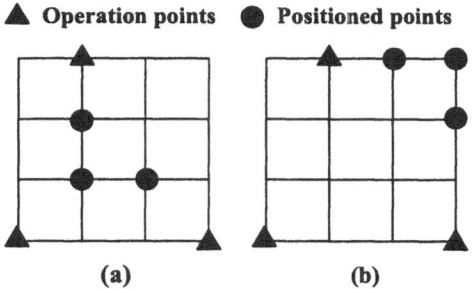

Figure 5: Feasibility of positioning operation

4 Control Of Indirect Simultaneous Positioning

4.1 Iterative Control Law

In this section, I propose a novel control method to achieve an indirect simultaneous positioning. An iterative control law is derived based on a linearized model of the object eq.(6).

In the control of indirect simultaneous positioning of deformable objects, a vision system is utilized to measure current positions of positioned points. Moreover, mechanical fingers pinch an object firmly so that no slip occurs between the fingers and the object at the operation points. This implies that the locations of operation points coincide with endpoints of the fingers, thus positions of the operation points can be computed from the locations of the endpoints. On the other hand, it is difficult to measure positions of non-target points due to their number.

Now, let us derive an iterative control law for indirect simultaneous positioning based on linearized equations (6). Fig.6 and 7 show a flow of our proposed control method and its schematic diagram, respectively. Assume that the number of positioned points is equal to the number of operation points, that is, $m = p$. In this case, F is a square matrix. This article deals with only the case that the matrix F_k is non-singular for any k for simplicity. Then, eq.(6) can be rewritten as the following two equations:

$$\delta r_m = -S_U F^{-1} C \delta r_p \tag{12}$$
$$\delta r_n = -S_L F^{-1} C \delta r_p \tag{13}$$

where

$$S_U = [I_m \ 0_{m \times n}] \tag{14}$$
$$S_L = [0_{n \times m} \ I_n]. \tag{15}$$

Let r_m^k, r_n^k and r_p^k be positions of operation points, those of non-target points, and those of positioned points at k-th iteration, respectively. Recall that the positions of r_p^k is measured by a vision sensor. Let r_p^d be a set of desired locations of positioned points. Let us derive an iterative law for updating desired locations of operation points based on eq.(12). In eq.(12), replacing deviation δr_m with difference $^dr_m^{k+1} - r_m^k$ and deviation δr_p with error $r_p^d - r_p^k$, we obtain the following equation:

$$^dr_m^{k+1} = r_m^k - dS_U F_k^{-1} C_k (r_p^d - r_p^k) \tag{16}$$

where F_k and C_k are functions of r_m^k, r_n^k, and r_p^k. Superscript and subscript k on variables denote their values at k-th iteration. A scalar d denotes a scaling factor. The right hand side of this equation can be evaluated at the k-th iteration. Thus, desired locations of operation points at the k-th iteration can be updated into those at the $(k+1)$-th iteration by this equation. Note that matrix F_k^{-1} depends not only r_m and r_p but also r_n. Thus, it is necessary to estimate locations of non-target points r_n. Let us derive a recursive law for estimating positions of non-target points based on eq.(13). Let us replace deviation δr_n with difference $r_n^k - r_n^{k-1}$. Note that deviation δr_p should be evaluated at the $(k-1)$-th iteration. These replacements yield the following equation:

$$r_n^k = r_n^{k-1} - dS_L F_{k-1}^{-1} C_{k-1} (r_p^{k-1} - r_p^{k-2}) \tag{17}$$

The right hand side of this equation can be evaluated at the $(k-1)$-th iteration. Thus, positions of the non-target points at the k-th iteration can be estimated

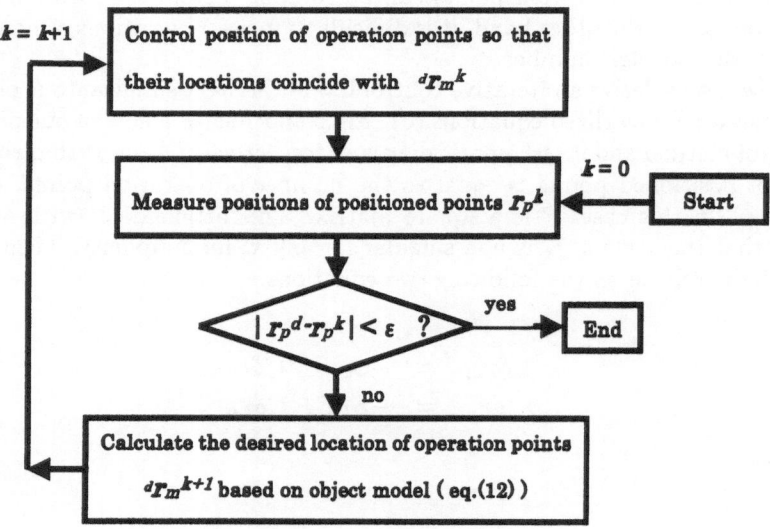

Figure 6: Flow of proposed control method

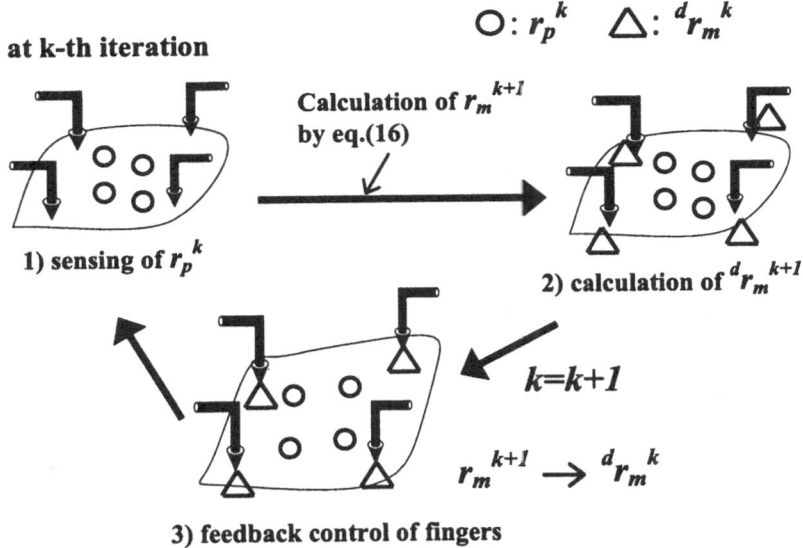

Figure 7: Schematic diagrams of proposed control method

As a result, the proposed iterative control method is summarized as follows: First, a vision system senses current positions of positioned points. Second, locations of operation points and those of non-target points are updated using eq.(16) and (17), respectively. Then, robot fingers are controlled with respect to task oriented coordinates using $^{d}r_m^{k+1}$ as their desired positions in $(k+1)$-th iteration with an appropriate controller. For example, we can utilize linear PID feedback. Asymptotic stability of r_m to $^{d}r_m^{k+1}$ is guaranteed with PID feedback based on Lyapunov stability theorem in the similar manner as a robot manipulator control [16]. However, due to lack of the room of paper, the detail is omitted here. After robot fingers converged to $^{d}r_m^{k+1}$, positions of positioned points r_p^{k+1} are measured again by the image sensor. Then, the same procedure is iterated.

4.2   Analysis Of Convergence To Desired Locations

Here, I will analyze the convergence of the positioned points to the desired positions.

In practice, we have to use the following approximated equation instead of eq.(16) because matrices include some errors:

$$r_m^{k+1} = r_m^k - \tilde{Q}_k(r_p^d - r_p^k), \tag{18}$$

$$\tilde{Q}_k = dS_U \tilde{F}_k^{-1}\tilde{C}_k, \tag{19}$$

where $\tilde{F}_k$ and $\tilde{C}_k$ denote estimated matrices of F_k and C_k, respectively and include some errors. The errors are mainly caused by identification errors of spring coefficients in the object model. The errors also come from the use of the linearlized model. Here, I consider the case that $r_m^{k+1} - r_m^k$ and $r_p^{k+1} - r_p^k$ are sufficiently small. Then, the following equations are satisfied according to eq.(6):

$$r_p^{k+1} - r_p^k = -Q_k^{-1}(r_m^{k+1} - r_m^k), \tag{20}$$

$$Q_k = S_U F_k C_k, \tag{21}$$

where there exists Q_k^{-1} because I assume that $\det F_k \neq 0$. Substituting eq.(18) into (20) yields

$$r_p^d - r_p^{k+1} = (I - Q_k^{-1}\tilde{Q}_k)(r_p^d - r_p^k). \tag{22}$$

Finally, we obtain

$$||r_p^d - r_p^{k+1}|| \leq ||I - Q_k^{-1}\tilde{Q}_k|| \, ||r_p^d - r_p^k||. \tag{23}$$

Therefore, if $\tilde{Q}_k$ satisfies

$$||I - Q_k^{-1}\tilde{Q}_k|| < 1, \tag{24}$$

we can guarantee

$$r_p^k \to r_p^d \quad as \quad k \to \infty. \tag{25}$$

Decreasing d yields the reduction of each element of estimated matrix $\tilde{Q}_k$. This implies that we can maintain the convergence to the desired positions by decreasing d despite large errors of matrix $\tilde{Q}_k$. The speed of convergence is, however, reduced when scaling factor d decreases.

5 Experiments

In this section, I will show experimental results in order to illustrate the validity of the proposed control method and to investigate the effect of model errors on the convergence quantitatively. Fig.8 illustrates the experimental setup. Three 2DOF robots with stepping motors are utilized as robotic fingers. A CCD camera is utilized as a vision sensor. A deformable object is laid on a table. In the experiments, knitted fabrics of the acrylic 85[%] and wool 15[%] (100[mm]×100[mm]) are utilized. The fabric is descritized into 4×4 meshes. Both of the numbers of the operation and positioned points are three. Their initial locations on the object are shown in Fig.9. Markers are put on the positioned points of the fabric. Their positions are measured by the CCD camera. The configurations of the operation and positioned points are as follows:

$$r_m = [x_{0,3}, y_{0,3}, \quad x_{1,0}, y_{1,0}, \quad x_{3,2}, y_{3,2}]^T,$$

$$r_p = [x_{1,1}, y_{1,1}, \quad x_{1,2}, y_{1,2}, \quad x_{2,2}, y_{2,2}]^T.$$

The desired positioned points used in the experiments are
$$r_p^d = [30, 40, \ 65, 50, \ 53.6, 90]^{\mathrm{T}}.$$

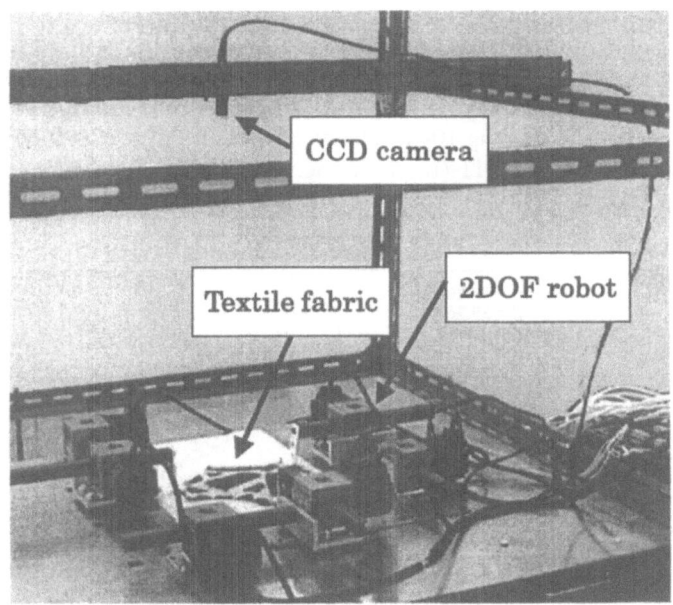

Figure 8: Experimental setup

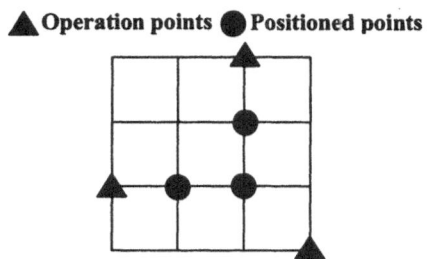

Figure 9: Configuration of points in experiments

The positioning tasks are executed with the method proposed in section 4. Namely, the positions of positioned points are measured by CCD camera. The measured positions are utilized for calculating the desired positions of the operation points using eq.(16) at each trial. Then, robot fingers grasping the operation points are controlled so that their positions converge to the desired ones. After the convergence, the positions of positioned points are measured again. The same procedure is iterated.

I have identified spring constants $(k_x, k_y, k_\theta) = (4.17, 13.2, 3.32)$ [gf/mm] coarsely for the control method, through tensile tests. Note that the ratio of the spring constants is important in our control method. Then, I define $\alpha = k_x/k_\theta$ and $\beta = k_y/k_\theta$. From coarsely identified spring constants, $\alpha = 1.256$ and $\beta = 3.976$ are obtained. In experiments, various values of α and β including errors were utilized in the control method, in order that the effects of the deviations of α and β from the identified ones are investigated. Moreover, values 0.1 and 0.5 of scaling factors d are used in eq.(16) to show that the effects of the scaling factors. Fig.10-(a) illustrates the experimental results of $d = 0.1$ and α is fixed. Fig.10-(b) shows the results of $d = 0.1$ and β is fixed. Similarly, Fig.11-(a) and 11-(b) show the results with $d = 0.5$, and α and β is fixed, respectively.

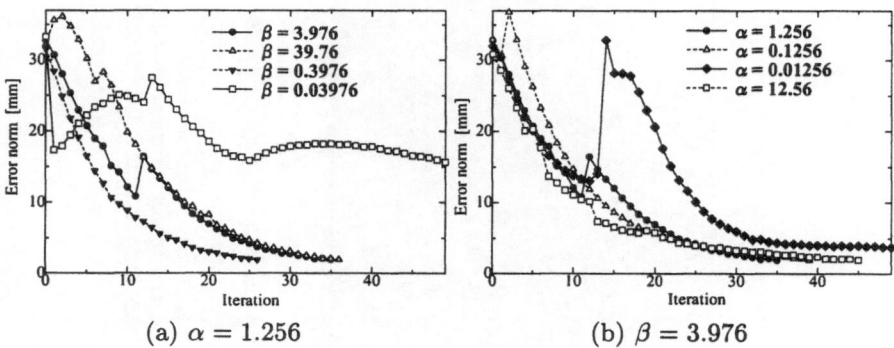

(a) $\alpha = 1.256$ (b) $\beta = 3.976$

Figure 10: Experimental results ($d = 0.1$)

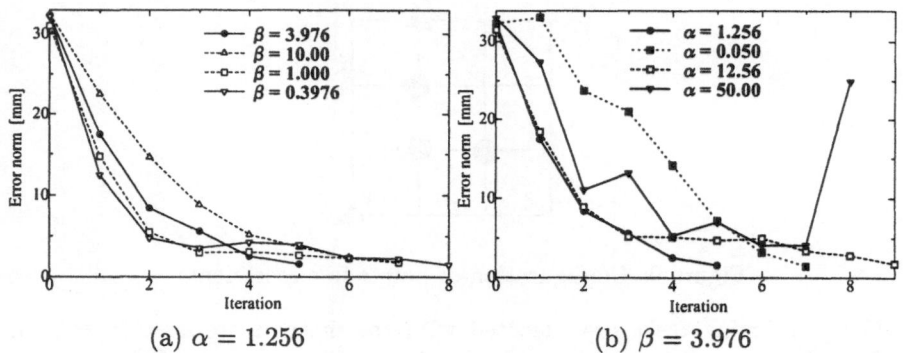

(a) $\alpha = 1.256$ (b) $\beta = 3.976$

Figure 11: Experimental results ($d = 0.5$)

In all of these figures, we can find that the positioned points can converge to the desired positions if the coefficients (α, β) is near their identified values, while they are gradually oscillatory or diverge if (α, β) is far from the identified values. Fig.10 shows that the positioned points converge to the desired ones despite of 100 times or 0.01 times of deviations of parameters α and β. Fig.11

shows that the positioned points diverge for 10 times and 0.1 times deviations of the parameters. On the other hand, the speed of convergence is higher with $d = 0.5$. As an example, Fig.12 shows behaviors of operation and positioned points with $d = 0.5$, $(\alpha, \beta) = (1.256, 3.976)$. The accuracy of convergence to the desired ones can be reached to a resolution level of the visual sensor (about 1[mm]).

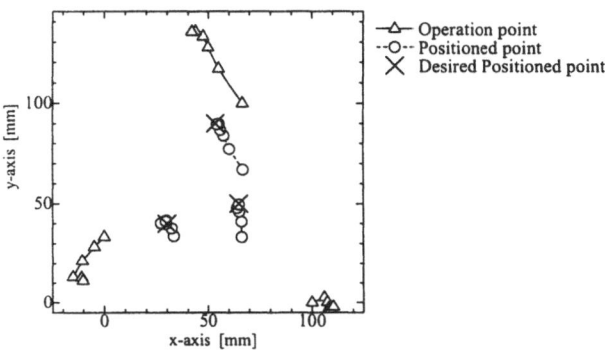

Figure 12: Behavior of positioned and operation points in experiments

According to the experimental results, we can conclude that very coarsely estimated parameters can be utilized in the proposed control method in a practical operations of the fabrics. Scaling factor d should be chosen carefully.

6   Conclusions

In this paper, indirect simultaneous positioning of deformable objects were discussed. First, I have proposed a coarse model of extensible deformable objects for their positioning operations. Second, indirect simultaneous positioning of deformable objects have been formulated. Based on the formulations, I analyzed the indirect positioning. As the result, I derived the conditions that the given positioning can be achieved. Then, I have proposed a novel iterative control method to realize indirect simultaneous positioning. The validity of the method has been shown through experimental results using the textile fabrics, and effects of the model errors on the convergence were investigated. Then, I conclude that very coarse identifications can be utilized for the proposed method.

In this article, I have proposed the control method to guide the positioned points to given desired location exactly for indirect simultaneous positioning. For these positioning, the same number of the operation points as that of the positioned points are needed. However, in practice, it is sometimes required that positioned points are guided to given desired regions. In these cases, it is expected that the number of the operation points are reduced in relative to that of positioned points. Also, initial locations of operation points on a deformable object were given before executing the task in this article. However, the task may be failed in or excessive forces may be exerted on the object in the case that the configuration of the positioned points is not appropriate. Then, task planning including configuration of operation points is important. Therefore, I have proposed the method to plan configurations of robot fingers on a deformable object so that positioning within desired regions can be achieved with appropriate magnitude of forces. The details have been reported in [17].

References

[1] Taylor, P.M. et al.(Ed.), "Sensory Robotics for the Handling of Limp Materials", Springer-Verlag, 1990

[2] Hall, M.K., Bill, B., and Bill L., "Methods for Automated Separation of Fabric Plies From a Cut Stack", Knitting International, pp.162–166, 1987

[3] Kemp, D.R., Taylor, G.E., Taylor, P.M. and Pugh, A., "A sensory gripper for handling textiles", in Pham, D.T.et al.(Eds.), "Robot Grippers", IFS Publication Limited, pp.155–164, 1986

[4] Ono, E. Ichijo, H., and Aisaka, N., "Flexible Robotic Hand for Handling Fabric Pieces in Garment Manufacture", International Journal of Clothing Science and Technology, vol.4, no.5, pp.16–23, 1992

[5] Abernathy, F., Pippins, D., Bray, F., Vento, V., "The Story of $(TC)^2$ Apparel Automation Research", Apparel International, pp.37–45, 1987

[6] Torgerson, E., Paul, F.W., "Vision Guided Robotic Fabric Manipulations for Apparel Manufacturing", Proc. of IEEE Int. Conf. on Robotics and Automation, pp.1196–1202, 1987

[7] Gershon, D., Porat, I., "Vision Servo Control of A Robotic Sewing System", Proc. of Int. Conf. on Robotics and Automation, pp.1830–1835, 1988

[8] Ono, E., Kita, N., Sakane, S., "Strategy for Unfolding a Fabric Piece by Coorperative Sensing of Touch and Vision", Proc. of Int. Conf. on Intelligent Robots and Systems, pp.441–445, 1995

[9] Hirai, S., Wakamatsu, H., and Iwata, K., "Modeling of Deformable Thin Parts for Their Manipulation", Proc. IEEE Int. Conf. on Robotics and Automation, pp.2955–2960, 1994

[10] Wakamtatsu, H., Hirai, S., Iwata, K., "Static Analysis of Deformable Object Grasping Based on Bounded Force Closure", Proc. IEEE Int. Conf. on Robotics and Automation, pp.3324-3329, 1996

[11] Howard, A.M. and Bekey, G.A., "Recursive Learning for Deformable Object Manipulation", Proc. of Int. Conf. on Advanced Robotics, pp.939–944, 1997.

[12] Sun, D., Liu, Y., Mills, J.K., "Cooperative Control of a Two-Manipulator System Handling a General Flexible Object", Proc. of Int. Conf. on Intelligent Robots and Systems, pp.5–10, 1997

[13] Wada, T., Hirai, S., Hirano, T., Kawamura, S., "Modeling of Plain Knitted Fabrics for Their Deformation Control", Proc. of IEEE Int. Conf. on Robotics and Automation, pp.1960–1965, 1997

[14] Wada, T., Hirai, S., Kawamura, S., "Planning and Control of Indirect Simultaneous Positioning Operation for Deformable Objects", Proc. of IEEE Int. Conf. on Robotics and Automation, pp.2572–2577, 1999

[15] Wada, T., Hirai, S., Kawamura, S., "Indirect Simultaneous Positioning Operations of Extensionally Deformable Objects", Proc. of Int. Conf. on Intelligent Robots and Systems, pp.1333–1338, 1998

[16] Arimoto, S., "Control Theory of Non-linear Mechanical Systems   A Passivity-based and Circuit-theoretic Approach", Oxford University Press, 1996

[17] Wada, T., Hirai, S., Kawamura, S., "Analysis and Planning of Indirect Simultaneous Positioning Operation for Deformable Objects", Proc. of Int. Conf. on Advanced Robotics, pp.141–146 1999

Section 3.2

A Hybrid Position / Force Approach to the Exploitation of Elasticity in Manipulation

B. J. McCarragher

Abstract. A unique contribution and motivation of this work is the realization that, when manipulating a flexible load, there may be advantages to exploiting the flexibility, rather than minimizing it. To exploit the flexibility of the manipulated object, a hybrid position /force control scheme is used. Hybrid position / force control considers the task space of the manipulator as subdivided into either force or position controlled directions. Force regulation is achieved with accommodation control, which makes the manipulator behave as an admittance (i.e. a force input / position output) map. This approach complements the natural impedance of the load, manifested in terms of force outputs provoked by a position input from the arm. The main results obtained are the global asymptotic stability of a proportional controller, and the local asymptotic stability of a PI controller. Two case studies illustrate the application of the method to de facto benchmark problems in the area of the manipulation of flexible materials. These case studies are the two-arm bending of sheet metal and the insertion of a slender, flexible beam into a hole with friction. The outcomes of the experiments show the simplicity and efficiency of the method in achieving task goals.

1 Introduction

Many industrial processes involve the manipulation of flexible parts, such as the assembly of car bodies, the bending of sheet metal and insertion tasks with flexible cooling pipes. Many issues exist surrounding the automation of these processes, including grasping the load, planning and controlling the task execution, and monitoring tasks. In this paper we concentrate on solving the planning and control problems which arise in the manipulation of flexible materials.

A unique contribution and motivation of this work is the realization that, when manipulating a flexible load, there may be advantages to exploiting the flexibility, rather than minimizing it. Indeed, some manipulations and assemblies of flexible loads can only be accomplished if the flexibility of the parts is used. In the particular class of manipulation tasks studied here, the main advantage is the extended elastic region provided by the flexible loads. This property allows us to use force control without the performance requirements necessary for rigid loads. Force control is used to compensate for the inaccuracies in path planning and execution.

To exploit the flexibility of the manipulated object, a hybrid position /force control scheme is used. Hybrid position / force control considers the task space of the manipulator as subdivided into either force or position controlled directions. The underlying design philosophy guiding the selection of controlled variables is the exploitation of the elastic properties of the part. This is achieved by purposeful maintenance of elastic deformations so that force control is easier to achieve than would be the case in the manipulation of stiff (rigid to the limit) loads. Only quasi-static properties of the loads are modelled, of which elasticity is the most important. Elasticity is modelled as a nonlinear stiffness appearing in all directions of a Cartesian coordinate system.

Force regulation is achieved with accommodation control, which makes the manipulator behave as an admittance (i.e. a force input / position output) map. This approach complements the natural impedance of the load, manifested in terms of force outputs provoked by a position input from the arm. The analysis of the force control is done in task-space. The main results obtained are the global asymptotic stability of a proportional controller, and the local asymptotic stability of a PI controller

Two case studies are used to illustrate the application of the method to de facto benchmark problems in the area of the manipulation of flexible materials. These case studies are the bending of sheet metal and the insertion of a slender, flexible beam into a hole with friction. The outcomes of the experiments show the simplicity and efficiency of the method in achieving task goals

In the literature, there have been three main approaches to the modelling of flexible materials. First is the use of distributed parameter dynamic models and full compensation for the control of position and vibration. Complex models of the flexible beam are central to most of these studies [3, 11]. Control papers address both single and two-link flexible arms. Book [2] presents a survey of the control methods used in this area up to 1993. A full model approach may be necessary when high-speed motions of heavy payloads are considered. However, tasks involving constrained operations on flexible parts (e.g. assembly) present higher frequencies for the modes of oscillation than those for unconstrained motions. For assembly, there is a larger range of motion velocities and accelerations at which the load can be operated before the oscillation modes become important. Thus, many assembly tasks can be treated with quasi-static models.

The second approach uses quasi-static models for the flexible load deformation with path planning methods derived from the geometry of the deformation. For example, [4, 5, 10] exploit various methods to approximate the deformed shape with sinusoidal functions and / or minimization methods on the potential energy function of a flexible beam. Other researchers have used quasi-static models with position control [1, 8, 9]. The main problem with this approach is the need for accurate models since position control cannot accommodate the build-up of internal forces within the flexible load.

The third approach uses quasi-static models, or simple lumped parameter models, and force control methods for the manipulation of flexible loads. For instance, Nguyen and Mills [11] use a position / force decoupling procedure to implement a hybrid controller. Only the rigid body mode of the load is modelled, so that the control cannot suppress low-frequency oscillations. Meer and Rock [7] use object impedance control to manipulate a planar object that is linearly elastic along one direction. They had difficulties assigning force references to be tracked, making impedance control difficult to use. Also the inverse dynamics of the flexible load are needed for the control.

The work presented in this paper belongs to the third approach. However, the hybrid position / force control developed here views the flexibility of the load as a benefit to be exploited. In further contrast, an explicit inverse dynamic model is not required for the force control. Instead, an accommodation control of the forces is used [13], which is shown to be stable for a range of control gains and material stiffnesses. The accommodation map gives best performance when it matches the load's elasticity. Thus, a model of the load is beneficial, but not strictly required. As we exploit the load's elastic properties, the deformations generate a force field around the deformed configuration. The accommodation law is best suited to elastic loads (or environments). As such, the control deteriorates as the stiffness grows close to a rigid contact situation. The avoidance of such situations is an integral part of our task design philosophy, since planning controlled deformations of the load provides a region of reduced stiffness to the force controller.

2. Modelling of the Flexible Part

For simple flexible parts such as beams and plates, it might be possible to obtain an analytical description of the geometry of the deformation. In the robotics literature, most flexible objects are modelled as flexible beams [3, 12]. These models generally result in higher-order or partial differential equations. While analytical models have the attraction of being mathematically robust, they are very difficult to use for practical problems. One is limited to a set of simple objects with consistent cross sectional areas. Additionally, in order for the model to be useful in operation, many parameters must be measured or observed. In many cases, measuring parameters of flexible materials is highly impractical.

Alternatively, potential energy methods [4] or finite element methods [5] are often proposed. Again, these methods offer a desirable mathematical rigor. However, they are very difficult to use in practice. Potential energy methods are applicable to general loads, but they require full knowledge of the boundary conditions on the flexible part. Additionally, the models become prohibitively complex as the parts themselves increase in complexity. Finite element methods (FEM) are very popular for product design purposes, but are difficult to incorporate into a controller design. Like potential energy methods, FEM requires knowledge of the boundary conditions of the flexible object. Furthermore, it is difficult with a control task to determine appropriate control variables and parameters to monitor with the complex models that both potential energy methods and FEM produce.

In this paper we take a non-linear model approximation backed up with experimental verification. Deformations of an elastic load create a region in which a relationship between force and position appears. The forces in this region are restoring in nature. Since every point within this region can be associated with a force vector, we say the elastic deformations create a force field in the workspace of the manipulator.

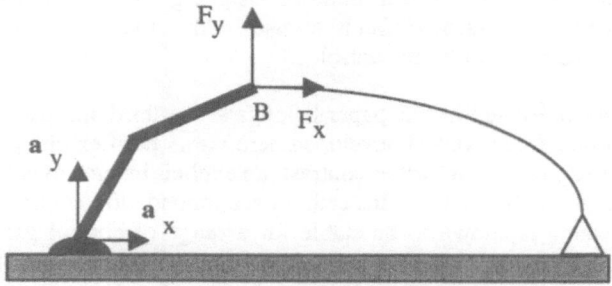

Figure 1: Bending a Flexible Beam without Wrist Torque

For motivation, consider a planar flexible beam under deformation, as shown in Figure 1. When the system is static, the force F exerted by the manipulator is solely a function of the end-point position B. Such position dependence can be represented as a force field around B:

$$F(x) = [F_x; F_y]$$

where $x = (x, y)$ is the position vector of B. In order to model the force x-position relation, a stiffness matrix is defined according to :

$$\kappa = \frac{\delta F_x}{\delta x}$$

The above equation expresses a differential relation between force and position. The sign of the elements indicate the direction in which the forces increase. For instance, an increase in y would cause a drop in F_x around the configuration shown in this figure. Thus, κ_{yy} is negative.

We make the following assumptions with regard to the stiffness parameters κ. First, κ is determined only within the elastic limits of the load, so the results are restricted to elastic region. Second, the force F must be continuously differentiable with respect to x, so that κ is well defined. As such, abrupt changes in the mode shape of deformation are not allowed.

Since κ is defined locally, a function of $\kappa(x)$ can be defined as the polynomial interpolation of the local values of κ. In fact, the experimental determination of κ leads naturally to a polynomial expression if interpolation of the local values is used. Hence, we define $\kappa(x)$ to be a polynomial of the form

$$\kappa(x) = \kappa_0 + \kappa_1 x + \ldots + \kappa_n x^n$$

For convenience, we wish to write F_x as a product between x and a stiffness-like coefficient. As such, we introduce a non-linear stiffness function $K(x)$, such that

$$K(x) = \kappa_0 + \frac{1}{2}\kappa_1 x + \ldots + \frac{1}{n+1}\kappa_n x^n$$

Using this definition of $K(x)$, we can express the local force field as

$$F_x = K(x)\, x - K(x_0)\, x_0$$

Or expanding to a general 6 degree-of-freedom case, the relation becomes

$$\begin{bmatrix} F \\ \tau \end{bmatrix} = \begin{bmatrix} K_{xx}(x\ \theta) & K_{\theta x}(x\ \theta) \\ K_{x\theta}(x\ \theta) & K_{\theta\theta}(x\ \theta) \end{bmatrix} \begin{bmatrix} x \\ \theta \end{bmatrix} - \begin{bmatrix} K_{xx}(x_0\ \theta_0) & K_{\theta x}(x_0\ \theta_0) \\ K_{x\theta}(x_0\ \theta_0) & K_{\theta\theta}(x_0\ \theta_0) \end{bmatrix} \begin{bmatrix} x_0 \\ \theta_0 \end{bmatrix}$$

As the K_{ii} elements are 3x3 sub-matrices, K can be defined as the nonlinear stiffness tensor following the nomenclature in elasticity theory.

We determine the stiffness tensor experimentally. This is done using small displacements δx and measuring the corresponding variation δF at many points in the workspace of the manipulator. The coefficients are then approximated according to

$$K_{ij} = \frac{\delta F_i}{\delta x_j}$$

An example set of results is shown in Figure 2. A 300 mm plastic ruler was brought into contact with a stiff environment. The orientation of the end-effector was constant. Variations in position were commanded, and the differences in forces were measured in order to determine the stiffness parameters. The nonlinear stiffness is then determined by integrating the interpolating polynomial of the local stiffness.

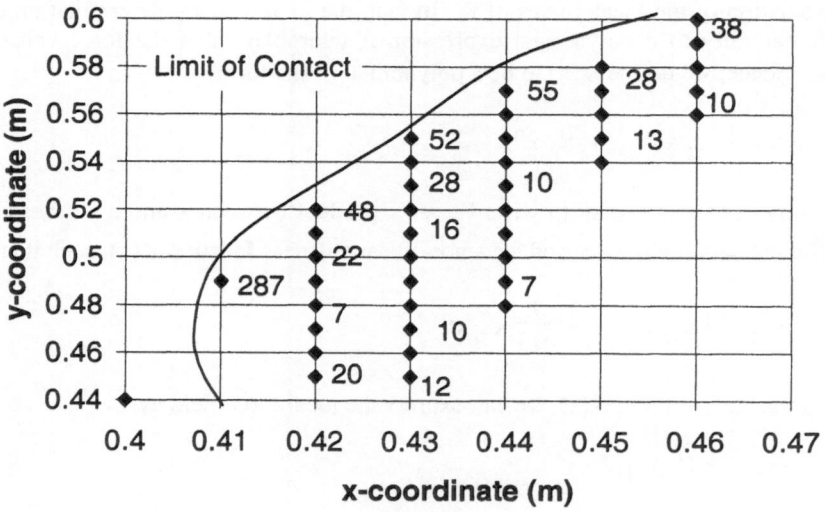

Figure 2: The local stiffness values in the horizontal direction

In the example above, F_x can be written as

$$F_x(x) = K_x(x)\, x - 411.6$$

where the nonlinear stiffness function is given by

$$K_x(x) = 3.33e10^3\, x^2 - 5.0e10^3\, x + 2.49e10^3$$

The advantages of an experimentally determined, non-linear stiffness model is that the method can be used for any deformable object. Additionally, the model can be made more accurate in areas of concern, and less accurate in areas that are not of concern. The disadvantage is that the method may lack the modelling elegance of, say FEM. However, the experimental method may be less cumbersome than some FEM models of complex parts.

3 Hybrid Position/Force Control

Given the experimental modelling of the flexible parts, we formulate the control using a hybrid position / force control framework. In this framework, the problem of synthesis of formal instructions reduces to the selection of variables to be controlled in each direction of the task space (either force or position). This allows for internal consistency of the instruction set because it avoids conflicting commands to be issued to the controllers. However, external consistency has to be verified according to the constraints imposed by the elements of the task. For instance, a set might be internally consistent by clearly separating the force and position controlled directions, but externally inconsistent by specifying a force-controlled direction in which the manipulator is unconstrained in position.

When flexible materials are involved, there is not unique solution to the problem of external consistency, because there are no clear-cut constraints in force or position. Rather, the constraints appear as force / position relationships, with the load's stiffness acting as a map between the two. The duality between force and position brings a distinctive advantage in that force control can be achieved with end-effector motions. This is clearly not possible in rigid contacts, since end-effector motions cannot take place. Rather, for rigid contacts, force regulation is achieved via reduced joint stiffness or direct torque control of the joint motors. By contrast, when the material is elastic, the joints can be made stiff (high control gains) for accurate tracking of angular references. In order to exploit the advantage of flexible materials, we define the following decision rule: The choice of command variables along a given direction of the task space should be such that elastic deformations are preserved on the load.

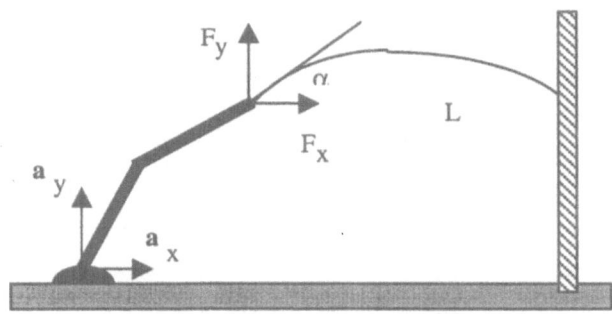

Figure 3: Bending a Flexible Beam without Wrist Torque

Consider the environment-following task depicted in Figure 3. A flexible beam of length L contacts the surface and follows it. Table 1 shows the natural constraints that are valid under the assumption of frictionless contact. K(.) is the nonlinear stiffness function. Planning and controlling the execution of a surface following task when the beam is straight and perpendicular to the surface is difficult, since any small motion away from the surface causes a loss in contact. With an elastically deformed beam, however, small motions away from the surface do not result in a loss of contact. Thus, it is to the designer's advantage to keep the beam deformed, since the surface following task easier to achieve. Effectively, the designer has the ability to decide, within the load's elastic limits, on a directional stiffness for the given task. In fact, by proper selection of the deformed state of the load along the execution path, the stiffness can be 'controlled' in certain directions.

Direction	Natural Constraint	Condition
a_x	$\dot{x} = 0$	$|F_x| < |F_x^{crit}|$ and $\alpha=0$, $x=1$
	$F_x = -K_x(x,\alpha)\, x$	All other cases
a_y	$F_y = 0$	----
a_z	$M_z = -K_z(x,\alpha)\, \alpha$	----

Table 1: Natural Constraints and Conditions for an Environment Following Task

We now move on to the problem of how to control the system in orthogonal directions as formulated in task space. Both force and position control loops generate velocity commands based on tracking errors of the respective variables. These commands are then transformed into commands at the joint level.

The most important type of force sensed by the manipulator is the restoring force created by the elastic deformation and gravitational forces. These forces are quasi-static, in contrast to the dynamics forces that would develop if oscillations were allowed on the load. From a control perspective, gravitational forces are imposed on the system. Thus the relevant force control problem refers to the regulation of forces generated by the elastic deformations of the load. We have selected an accommodation control [13], which issues a velocity command in response to an error in the force signal. Accommodation control is well suited to our framework because there are regions of reduced stiffness created by elastic deformations of the flexible loads. The novelty of its use in our context comes from the nonlinear nature of the load's stiffness and how this stiffness is incorporated into the control law.

Using the models developed previously, the restoring force exerted by the load on the manipulator is written as

$$\mathbf{F} = \mathbf{K}(.)\mathbf{x}$$

where K is the nonlinear stiffness of function for the load. We assume that the restoring forces always increases in the direction of increased deformation.

$$\kappa = \frac{\delta F_x}{\delta x} > 0$$

As such, the force control law is based on a velocity command so that the manipulator's end-effector moves in a direction that decreases the force error.

$$\dot{x}_r = b \, \Delta F$$

where ΔF is the difference between the measured force and the reference force, and b is the damping constant. The following lemma is proved in [6]

> Lemma: If the velocity command can be perfectly tracked under the influence of the restoring force, then the force error ΔF decreases monotonically to zero for a constant force reference

The lemma provides further rationale for the selection of the accommodation law.

The convergence rate is basically defined by the accommodation gain, b, subject to stability conditions imposed by the dynamics of the manipulator system (see [6] for a full discussion of the stability analysis of the force control scheme).

4.0 Experiments with Flexible Parts

We present two experiments that have been implemented using the hybrid position / force framework described above. These two experiments have been extensively studied in the literature of robotic manipulation of flexible materials. The first task is the bending of a flexible beam using two robot manipulators [1, 5, 11, 14]. In this case, the load deformations are an integral part of the task's requirements. By contrast, the second task is the insertion of a flexible beam inside a hole with friction [8, 9]. In this case, there is no requirement for deformation. However, elastic deformations are introduced so that task execution is more effective.

Another distinction is the accuracy of the geometric models for the tasks. For bending, analytical models allow for precise planning and analysis of the strategies. In the case of the insertion task, empirical reasoning is used since neither analytical nor numerical models can be obtained due to unknown boundary conditions at the inserted end of the flexible part.

4.1 Bending of a Flexible Beam

In the bending of a flexible beam with two manipulators, the coordination of the two arms is of central importance. Coordination of position controlled directions in a hybrid position / force controller is an off-line issue. Care must be taken so that appropriate position references are simultaneously issued to the arms, which can be decided off-line. To compensate for the build up of internal forces, however, the arms must also be controlled in force. The introduction of force control strongly couples the arms through the internal forces, with actions of one arm reflecting immediately on the other. This dependency affects the trajectories of the whole system and, as such, coordination becomes an on-line concern.

The coordination strategy proposed here is based on the examination of the equilibrium conditions in every force controlled direction of the ask space. However, force control cannot be used for the two arms simultaneously because a drift may occur [6]. As such, we use a mixed master / slave approach, whereby the master arm is controlled in position while the other follows according to the force requirements for optimality.

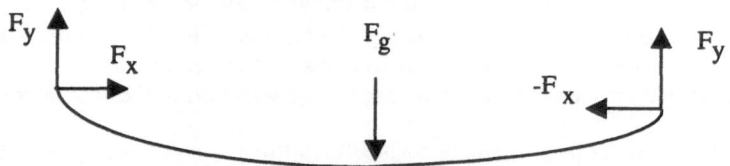

Figure 4: Beam under general loading for two arm bending task

The bending task is depicted in Figure 4. We present a strategy consisting of two phases. Initially, circular bending [6, 15] is deployed. Circular bending consists of a rotation applied by the end effector while controlling the forces to be zero. This action is equivalent to a pure-torque bending, since the only applied bending effort is a torque in the a_z direction. The circular bending phase is used to insure that the beam is deformed before any further bending occurs. If the beam is straight, compression in the a_x direction is possible, leading to a build-up of forces and to a potentially damaging abrupt change in the a_x beam configuration when the force went above the critical load. For our example, the commands for implementing circular bending are given in Table 2. Under these commands, the beam assumes a circular-arc shape, hence the name circular bending.

Direction	Controlled Variable	Left Arm Command	Right Arm Command
a_x	Master / Slave	$x_r = x_0 + \dfrac{1}{2}\left[1 - \dfrac{\sin(\alpha_r)}{\alpha_r} \right]$	$F_{xr} = 0$
a_y	Master / Slave	$y_r = y_0$	$F_{yr} = \dfrac{F_g}{2}$
a_z	Position	$\dot{\alpha}_r = C_1$	$\dot{\alpha}_r = -C_1$

Table 2: Commands for Phase 1 Circular Bending of Beam with Two Arms

Once an initial bending angle has been achieved using circular bending, we apply *elastica* bending. Previous work [6, 14, 15] has shown that elastica bending minimizes the effort applied by the end-effector. Elastica bending consists of a motion along the horizontal, while the vertical force F_y and the moment M_z are controlled to setpoints. Again we use the hybrid position / force control scheme developed previously. The vertical force setpoint is chosen to cancel the force of gravity $-\dfrac{F_g}{2}$ on each arm. The moment setpoint is chosen to be zero, so that there are no moments acting on the beam. As such, the commands for the elastica bending are given in Table 3

Direction	Controlled Variable	Left Arm Command	Right Arm Command
a_x	Position	$\dot{x}_r = C_2$	$\dot{x}_r = -C_2$
a_y	Master / Slave	$y_r = y_0$	$F_{yr} = \dfrac{F_g}{2}$
a_z	Force	$M_{zr} = 0$	$M_{zr} = 0$

Table 3: Commands for Phase 2 Elastica Bending of Beam with Two Arms

In both phases the force commands are chosen so that optimality in terms of minimum effort is achieved. In Phase 1, this means a zero horizontal force component, and in Phase 2, a zero torque component. The vertical component of force acts only to balance the weight of the flexible beam. The position-controlled directions dictate the velocity of bending. In both phases, the angular velocities for circular bending have to be of opposite signs and the same magnitude. These commands guarantee that the arms are coordinated along those directions controlled in position. The coordination of the force-controlled directions is guaranteed by using a master/slave system in order to avoid a positional drift in the manipulators.

The implementation of this task was done using two Eshed Scorbot II arms with wrist mounted JR3 force sensors. PID controllers were used at the joint level. The controllers were implemented using VME boards with Motorola 68040 microprocessors running VxWorks in real time. The flexible beam was made of steel, with a length of 605mm, a width of 30mm, and a thickness of 1mm. More information on the physical experiments can be found in [6].

Figures 5 and 6 show the horizontal and vertical force components during the task. Note that these are sensed forces, so they represent the reaction of the flexible load to the forces applied by the arms. The two phases of bending are distinguished by the different force setpoints. For $0 < t < 10s$, circular bending requires a zero force in the horizontal direction. After $t > 10s$, elastica bending is executed, which takes the horizontal force to around -15 N. It is interesting to note that the horizontal force component seems to quickly reach equilibrium. However, a constant velocity in a_x should results in a steady increase in the corresponding component of the force. Actually, a 3% (approximately) increase in the horizontal force is needed for increasing the bending angle from –20 to –35 degrees. This small and steady increase in force is indistinguishable from the control oscillations and the noise in the force measurement. The vertical forces are controlled around $Fg/2 = -0.86$ N.

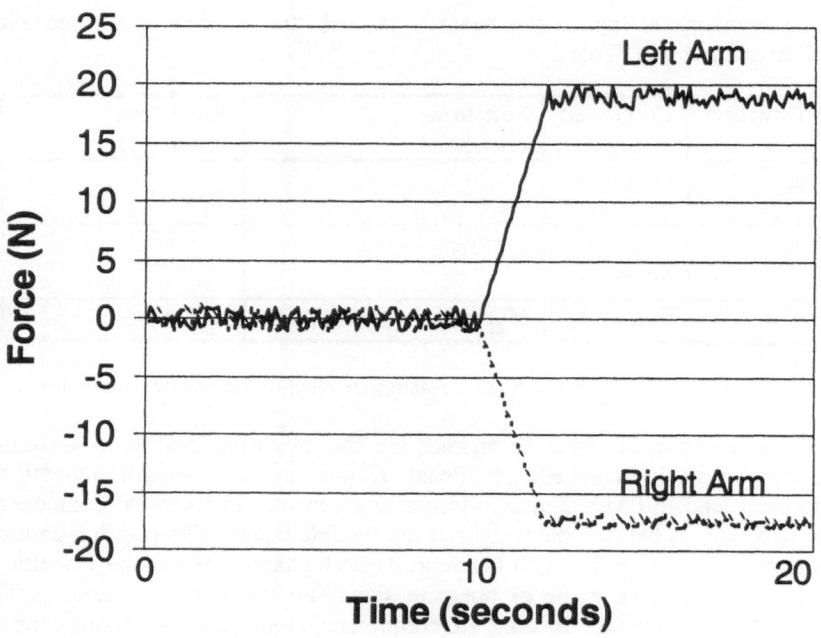

Figure 5: Horizontal force profiles for the two-arm bending task

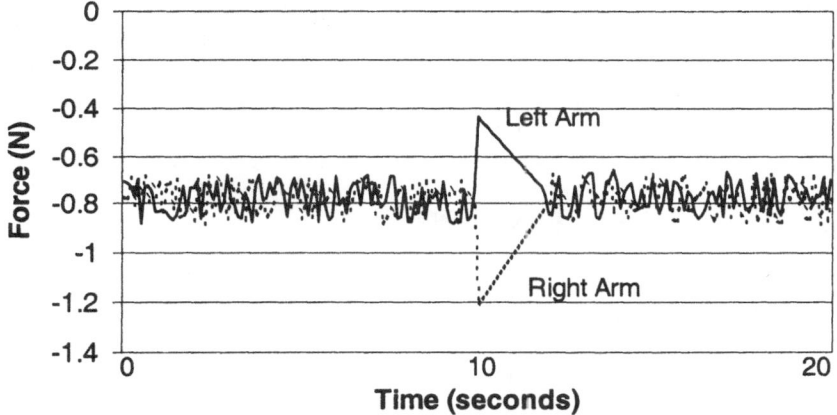

Figure 6: Vertical force profiles for the two-arm bending task

The torques applied to the ends of the beam are shown in Figure 7. The constant rotation applied to the arms during circular bending results in a steady increase in the torques. A drop in the applied torques to zero as required by the optimality formulation characterizes the transition to elastica bending. Note that the mixed master / slave scheme works well in controlling the torques at the zero setpoint.

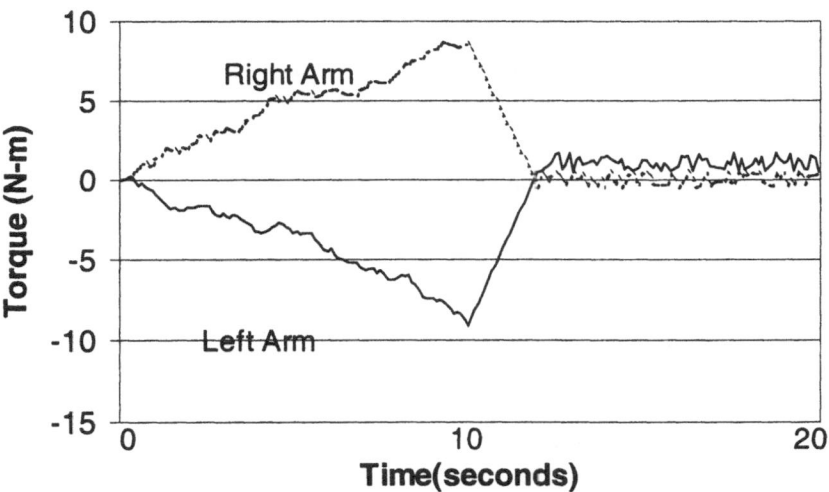

Figure 7: Wrist torques during the two arm bending experiment

During the successful experiment, the two arms each cover approximately the same distances during bending. The overall displacement in a_x is 25mm for the master (left) arm and 26mm for the slave. The displacements in $\mathbf{a}_y$ are on the range of +/- 0.1mm for the master arm and +/- 0.3mm for the slave. The latter moves more during circular bending due to the different nonlinear stiffness it experiences in that stage. The overall motions of the arms are comparable, indicating a good distribution of task-level effort. This observation is confirmed by the force profiles, which show that both arms are sharing evenly the bending effort.

4.2   Insertion of a Flexible Beam

The insertion task is conceptually different from the bending task. In bending, a geometric model of the load guided the choice of controlled directions and corresponding position and force commands. With the insertion task, however, a model of the load deformation is not available because the boundary conditions on the load are unknown and all three components of the force acting on the load are non-zero. Thus, task planning is done using a heuristic perspective.

In principle, no deformations are required on the load for insertion. However, deformations are purposefully included in order to exploit the elasticity. The deformation achieves three benefits. First, the restoring force generated by the deformation creates a firm contact between the tip of the load and the hole. Second, the deformation creates a region of reduced stiffness that allows for improved force control performance when compared to high-stiffness contacts. Third, the restoring force helps break down static friction so that insertion can proceed. Nonetheless, the control design is somewhat more involved due to the need for experimental determination of force references.

The flexible beam is similar to the bending task. The hole is a horizontal slot with a clearance of 2mm. The hole is such that friction creates a resistance in the direction of insertion. Friction increases with an increase in insertion depth, since the area of the ruler in contact increases. Initially, this longitudinal friction is small, and the main problem is the static friction arising from the contact between the tip of the beam and the hole, as shown in Figure 8. The main objective then is to break down this initial static friction, and proceed with insertion such that the longitudinal friction is kept to a minimum.

We again propose a two-phase approach to the insertion problem. The first phase seeks to progressively align the end effector with the hole, in a strategy similar to [8, 9]. A downward velocity is commanded in $\mathbf{a}_y$, while F_x is controlled so that contact between the tip of the beam in the hole is maintained. Critical to this contact maintenance is the selection of force references. In this case, the force references are determined empirically since analytical models are not available. For our experiments, the control commands for phase 1 are given in Table 4.

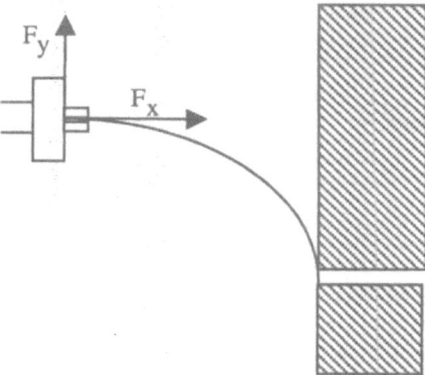

Figure 8: Inserting a flexible beam into a hole with friction

Direction	Controlled Variable	Control Command	Control Parameters
a_x	Force	$\dot{x}_r = b_x\,(F_{xr} - F_x)$	$b_x = 0.025$ m/(Ns)
a_y	Position	$\dot{y}_r = k_y\,(y_r - y) + C_1$	$k_y = 1\ s^{-1},\, C_1 = 5$ mm/s
a_z	Position	$\dot{\alpha}_r = k_\alpha\,(0 - \alpha)$	$k_\alpha = 1\ s^{-1}$

Table 4: Commands for Phase 1 of Flexible Insertion Task

When the static friction is overcome, phase 2 begins. This event is identified by a near zero value of the vertical force component. When this component is near zero, there is minimal force transversal to the axis of the hole, so that friction is minimized and insertion can proceed. The drop of the vertical force to near zero happens because (i) the end effector reaches the height of the hole; or (ii) the tip of the bean slips inside the hole when the ratio of the horizontal to the vertical forces becomes larger than the maximum static friction coefficient for the contact. Both occurrences have been observed in the experiments.

Once the beam is aligned with the hole and the static friction is overcome, phase 2 proceeds with forward motion. In order to keep friction to a minimum, the vertical force component is controlled to be near zero. The control design follows the pattern already used in the bending tasks. The accommodation and CLIK parameters are found experimentally. The commands for phase 2 are given in Table 5.

Direction	Controlled Variable	Control Command	Control Parameters
a_x	Position	$\dot{x}_r = k_x\,(x_r - x) + C_2$	$k_x = 1\ s^{-1},\ C_2 = 5\ mm/s$
a_y	Force	$\dot{y}_r = b_y\,(0 - F_y)$	$b_y = 0.1\ m/(Ns)$
a_z	Position	$\dot{\alpha}_r = k_\alpha\,(0 - \alpha)$	$k_\alpha = 1\ s^{-1}$

Table 5: Commands for Phase 1 of Flexible Insertion Task

The resulting force profile during an insertion task is shown in Figure 9. The insertion proved to be successful, although the system came close to losing contact, as evidenced in Figure 10. The x-y trajectory moves very close to the limit of contact. If the oscillations at this point were larger, a loss of contact could have occurred. To overcome this problem, either a stronger initial force bias or an increasing force reference is needed. A force reference as a function of degree of force oscillation (as shown in Figure 9) would be preferable. In this case, the control references are then reacting to the actual state of the flexible load. A robust method for monitoring the flexible load using just force sensing is still and open problem.

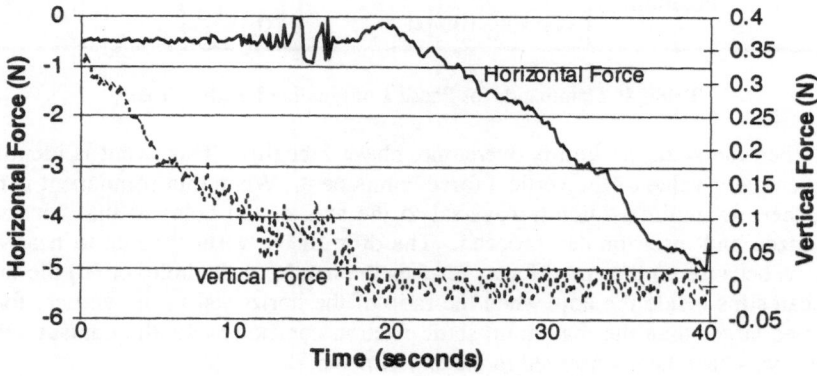

Figure 9: Force components during insertion task

Another interesting feature of the plots is the force oscillations prior to phase 2. The oscillations appear because tracking of the force reference backs the end-effector away from the hole during the downward motion. This backing leads to a region of low deformation for the beam. Figure 10 shows the path in x-y coordinates and the limit of contact. The limit of contact represents a line in the workspace for which f0 is zero. If the end effector moves past this line, the tip of the beam loses

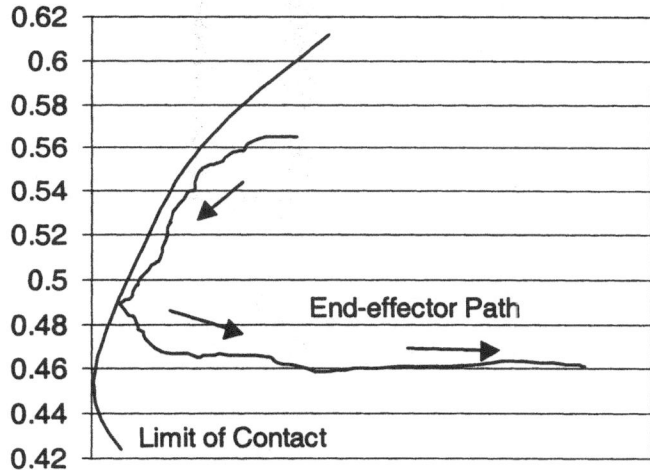

Figure 10: End-effector trajectory during insertion task

contact with the bottom of the hole. Again, this problem can be overcome through the use of a larger initial force reference or a force reference that is a function of the actual state of the system, as indicated by the force oscillations.

Lastly, one can see in phase 2 the decreasing value of force in the x-direction, indicating the increase in friction as the beam is inserted into the hole. Also, the force in the y-direction is kept nearly zero to ease insertion.

These two experimental case studies demonstrate the merits of our hybrid position / force control method. The bending experiment showed the control method applied to a system with a relatively good model of the flexibility of the load. The bending experiment also showed the use of a master / slave control framework incorporated within the hybrid control. A two-phase control approach was used to overcome some of dangers associated with the buckling of flexible materials. This approach allowed a simple and effective solution to a problem that can be quite complicated if force control is not used. The insertion experiment demonstrated effective control when representative models of the load deformations are not available. Again, a two-phase hybrid position / force control methodology was used. Due to the lack of models, the force commands were determined using an iterative, empirical method. Although not fully general, our approach to the insertion task yields one of the best experimental results reported in the literature [cf. 8, 9, 15].

5. Conclusion

The hybrid position / force control method developed here unifies, at the task control level, the solution to different manipulation problems involving flexible materials. The unifying element in our approach is the controller, which uses accommodation control of forces to actively exploit the reduced stiffness brought about by elasticity. A systematic procedure for task planning in a hybrid position / force control framework was also developed. The main novelty of the planning approach is the deliberate introduction of elastic deformations on the load, departing from more traditional approaches that avoid deformations. Also, empirical determination of force references is included as part of the planning and control method whenever models for the load cannot be obtained. Two sets of experiments were provided which demonstrated the effectiveness of the approach, particularly exploiting the flexibility of the materials. One goal of this work was the application in industrial settings. Our methods, as demonstrated by the experiments, are simple and robust enough to be deployed in present day industrial settings.

References

[1] Al-Jarrah, O, Y.F. Zhang, and K.Y. Yi. Efficient trajectory planning for two manipulators to deform flexible materials with experiments. IEEE Int. Conf. on Robotics and Automation, 1:312-317, May 1995.

[2] Book, W, Controlled motion in an elastic world. AMSE Journal of Dynamic Systems, Measurement and Control, 115(6):252-261, June 1993.

[3] Cetinkunt, S, and W.L. Yu, Accuracy of finite dimensional dynamic models of flexible manipulators for controller design. Journal of Robotic Systems 9(2):327-350, March 1992.

[4] Hirai, S., H. Wakamatsu and K. Iwata. Modelling of deformable thin parts for their manipulation. IEEE Int. Conf on Robotics and Automation, p. 2955-2960, April 1994.

[5] Kosuge, K., M. Saki, K. Kanitani, H. Yoshida, and T. Fukuda. Control of deformation of sheet metal by dual manipulators. Proc. of 3rd European Control Conference, p. 318-323, Sept. 1995.

[6] Kraus, Werner, The Robotic Manipulation of Flexible Materials: A Hybrid Position / Force Approach. PhD Thesis, Department of Engineering, Australian National University, March 1997.

[7] Meer, D.W. and S.M. Rock. Experiments in object impedance control for flexible objects. IEEE Int. Conf on Robotics and Automation, 2:1222-1227, 1994.

[8] Nakagaki, H, K Kitagaki and H Tsukune, Study of insertion task of a flexible wire into a hole. IEEE Int. Conf. on Robotics and Automation, 1:330-335, April 1995.

[9] Nakagaki, H, K. Kitagaki, T. Ogasawara, and H. Tsukune. Study of insertion of a flexible wire into a hole by using visual tracking observed by stereo vision. IEEE Int. Conf. on Robotics and Automation, 4:3209-3215, April 1996.

[10] Newell, G. C. and K. Khodabandehloo. Modelling flexible sheets for automatic handling and layup of composite components. IMechE Journal of Engineering Manufacture, 209:423-432, 1995.

[11] Nguyen, W. and J.K. Mills. Multi-robot control for flexible fixtureless assembly of flexible sheet metal auto body parts. IEEE Int. Conf. on Robotics and Automation, 3:2340-2345, April 1996.

[12] Pota, H.R. and T.E. Alberts. Multivariable transfer functions for a slewing piezoelectric laminate beam, ASME Journal of Dynamic Systems, Measurement and Control, 117(3):352-359, March 1995.

[13] Schimmels, J.M. and M.A. Peshkin. Admittance matrix design for force guided assembly. IEEE Trans. on Robotics and Automation, 8(2):213-227, 1992.

[14] Zheng, Y. F. and M.Z. Chen, Trajectory planning for two manipulators to deform flexible beams. Robotics and Autonomous Systems, 12:55-67, 1994.

[15] Zheng, Y.F., R Pei and M.Z. Chen, Strategies for automatic assembly of deformable objects. IEEE Int. Conf. on Robotics and Automation, p. 2598-2603, 1991.

Section 3.3

Force- and Vision-Based Detection of Contact State Transitions

F. Abegg, A. Remde, and D. Henrich

Abstract. A new and systematic basic approach to force- and vision-based robot manipulation of deformable (non-rigid) linear objects is introduced. This approach reduces the computational needs by using a simple state-oriented model of the objects. These states describe the relation between the deformable and rigid obstacles, and are derived from the object image and its features. We give an enumeration of possible contact states and discuss the main characteristics of each state. We investigate the performance of robust transitions between the contact states and derive criteria and conditions for each of the states and for two sensor systems, i.e. a vision sensor and a force/torque sensor. This results in a new and task-independent approach in regarding the handling of deformable objects and in a sensor-based implementation of manipulation primitives for industrial robots. Thus, the usage of sensor processing is an appropriate solution for our problem. Finally, we apply the concept of contact states and state transitions to the description of a typical assembly task. Experimental results show the feasibility of our approach: A robot performs several contact state transitions which can be combined for solving a more complex task.

1. Introduction

The manipulation of rigid objects by robots has been the subject of study for several decades. Less effort has been made in investigating the manipulation of non-rigid or *deformable* objects, despite its significance in many industrial applications [1]. Here, we focus on the one-dimensional or *linear* deformable objects, such as cables, wires, ropes, strings, beams, etc. This task has various application fields, e.g., hot-wire maintenance [2], cable form assembly, and production of control cabinets. The main problem of manipulating these objects is that they may change their shape during manipulation.

To cope with this problem, one approach is to estimate the shape of the deformable objects by calculating an internal model and simulating the object behavior. A static model of the objects and the obstacles can be calculated in two [3] or three [4] dimensions. An extension leads to a dynamic model of deformable linear objects [5]. On the one hand, the object shape can be calculated with these methods precisely (*direct* simulation problem). On the other hand, it is not clear how to use the object models to control the robot motion, that is, to solve the *inverse* simulation problem [6]. Additionally, the shape calculation can be very time consuming.

Another approach is to employ sensor systems to detect the object's shape. Vision systems can be used, for example, to guide the robot motion while the robot is making a knot with a rope [7], or to detect the shape of a flexible beam while inserting it into a hole [8, 9]. Similar to model-based approaches, the vision-based approaches are quantitative ways to measure or calculate the shape. Force/torque sensors measure forces acting on the deformable object and can be used to detect the buckling of the object when being inserted into a hole [9, 10].

All known approaches to vision-guided handling of deformable linear objects so far only present a solution for one special problem. In [11, 9, 12], a flexible beam is inserted into a hole. Nakagaki et al. additionally integrate a force/torque sensor while Chen et al. only use off-line sensor processing. In [7], a highly plastically deformable rope is used which is always hanging down and not deformed by a contact with an obstacle. Byun and Nagata compute the 3D pose of deformable objects but they have to deal with the stereo correspondence problem since they use a stereo vision approach [1]. In the work of Smith, highly plastically deformable ropes are laid along a desired shape in a plane [13], but non-elastic materials like ropes are not regarded in this work so far.

A qualitative sensor-based approach to manipulate deformable linear objects is skill-based manipulation. *Manipulation skills* are motion primitives for achieving a particular target state of the manipulated object. They are specified in the task domain independently of the robot hardware, and hide control procedures and sensor feedback from the programmer. Skills are robust and overcome residual errors and uncertainties in both, models and manipulator movements. For example, for rigid polyhedral objects, the manipulation skills serve as transitions between contact states, and they simplify programming for a model-based manipulation system [14]. The sequence of manipulation skills can then be extracted automatically from the motion performed by an operator in a simulator [15].

The basic precondition of skill-based manipulation is the identification of object states. Then, the manipulation skills can serve as transitions between these states. The question is what kind of state models are appropriate for deformable objects. Topological states, such as provided by the knot theory, use the number and kind of crossings of the linear object (with itself) [16]. For rigid objects, contact states differ in the involved type and number of geometric primitives [17]. Shape states are determined by calculating the precise or approximate object geometry [4, 18] and can hardly be distinguished in a symbolic way. Position states use the location and orientation of geometric primitives relative to other geometric primitives [19].

In this chapter, we formulate a general approach for the sensor-based manipulation of deformable linear objects (DLO). We investigate contact states and point contacts of deformable linear objects and show how our approach is used to implement task-independent manipulation primitives for industrial robots. These primitives are also referred to as skills and can be combined for solving different tasks.

For solving manipulation tasks with deformable objects, we need to answer the following questions: what are the possible contact states of deformable linear objects and how do deformable linear objects in contact behave qualitatively? How does a human recognize contact state transitions? How can the approach be used together with a sensor processing system for guiding a robotic handling system by observing the changes of the object shape? What are the results of using our approach? What are the conclusions and how should the work be continued and improved?

2. Contact State Transitions

We introduce contact states and contact state transitions of deformable linear objects with respect to an obstacle environment. Our approach of sensor-based operations bases on the recognition of them.

2.1 Contact states

In the following, the contact of a deformable linear object (called workpiece) in a static environment (called obstacle) is regarded. We develop the approach based on the following two assumptions:

The first, the material of the workpiece is isotrop and homogeneous. The workpiece is assumed to be *uniformly curved*, that is, it is either uniformly convex or concave. The deformation caused by gravity and contact forces is elastic, that is, the deformation disappears when the stress is released.[1] Example workpieces are a (short) hose or a piece of spring steal. The linear workpiece is gripped at one end and the robot gripper may perform arbitrary linear motions.

The second, all obstacles consist of convex polyhedrons. The friction between workpiece and obstacle is negligibly low. We begin our consideration with a single contact between workpiece and obstacle.

Based on the geometric primitives of DLO and obstacle, a set of contact states that enumerates all possible contact situations is derived [20]. For polyhedral

---

[1]   The workpieces belong to the object classes {E−, E+} introduced by [20].

objects, the geometric primitives include vertices (V), edges (E), and faces (F). The linear workpiece has two vertices and one edge between the two vertices. We name the contact states by the contact primitive of the workpiece followed by the contact primitive of the obstacle, for example V/F for vertex-face contact (Figure 1). An additional state is N which indicates that workpiece and obstacle are not in contact.

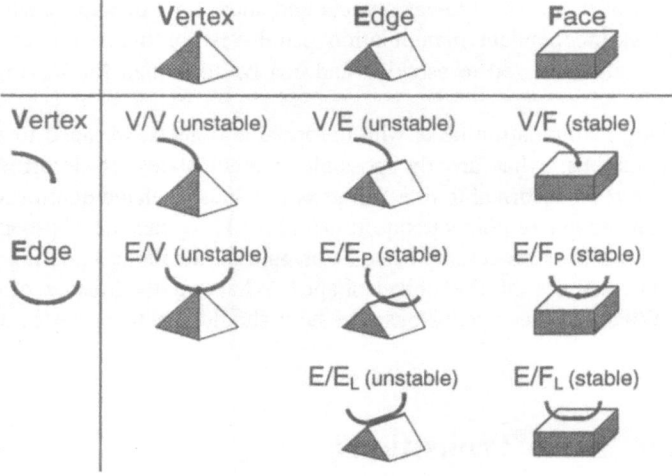

Figure 1: Enumeration of contact states between a deformable linear object and a convex polyhedron

An important attribute of each contact state is its stability. A contact state which remains unchanged as the robot gripper makes a (small) motion in any direction is called *stable*. However, the contact point or contact line may move. If this condition is not fulfilled, we call the contact state *unstable*. Consequently, a stable contact state is especially kept up when the robot gripper is not moved.

2.2 State transitions

State transitions are a change from one contact state to another without passing intermediate states. For now, establishing a second contact without loosening the first one, i.e., establishing a double contact, is not considered. Combining the contact states with the transitions between them, the graph shown in Figure 2 is obtained. This graph gives all possible transitions between the contact states (including state N) and is found by means of basic manipulation experiments. The contact states represent nodes while the state transitions represent edges with initiated transitions (plain arrows) and spontaneous transitions (dashed arrows). Solid edges starting and ending at the same node indicate stable contact states, i.e., these states can be their own successors. This is also the case when a motion of the robot gripper is not large enough to cause a state transition.

As stated in [21], any stable state may be directly established from state N. Stable states are connected with state N by solid edges. Transitions between stable states are reversible by just performing the same gripper motion in the reverse direction. For transitions beginning with unstable states, things are different. It is found that they are only partly reversible. The dashed edges in the transition graph starting from these contact states indicate that there are several possible stable successors for each of them. While transitions leading to a stable state different from N (e.g. V/F→V/E→V/F) are reversible, those transitions leading to N are irreversible [21].

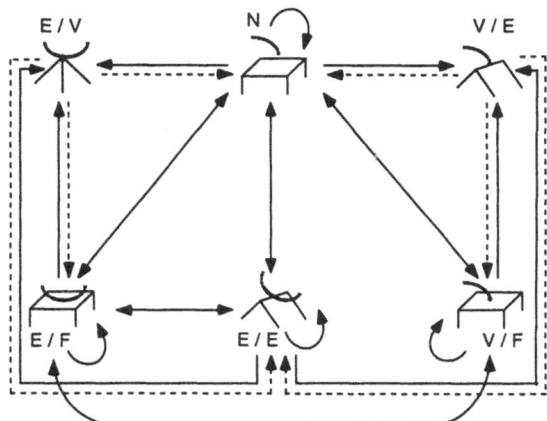

Figure 2: Contact state transition graph [21] with initiated transitions (plain arrows) and spontaneous transitions (dashed arrows)

2.3 Programming of manipulation tasks

So far, we have presented contacts of a deformable linear workpiece with a rigid obstacle and contact state transitions from an analytical point of view. This section introduces our concept to the application of assembly tasks.

As far as rigid workpieces are concerned, a lot of works address the problem of developing robust and flexible routines for typical assembling or disassembling tasks. The basic idea is to set up a library of encapsulated, sensor-based routines that can be used as a construction kit for efficiently solving complex assembly problems. Morrow and Khosla demonstrate the efficiency of this method in inserting different kinds of plug-in connectors [19].

Morris shows that performing assembly tasks can be regarded as a stepwise increasing of the number of constraints (reducing the degrees of freedom) of one of the mating parts by establishing contact with the other part [22]. Therefore, detecting and manipulating the contact state of the mating parts is a key issue for developing manipulation routines. Any routine that changes the contact state of the mating parts, like establishing point contact, transferring point contact to face contact, etc., forms a module of the construction kit for assembly operations.

When thinking of those tasks, it is found that the basic assumptions on our working environment must be partly relaxed. This affects especially the assumptions of one single contact and of a convex polyhedral obstacle.[2] However, this does neither affect the applicability of the set of contact states nor the rules derived for state transitions.

Let us consider the problem of inserting an elastic pneumatic hose into an U-shaped guiding groove as shown in Figure 3.

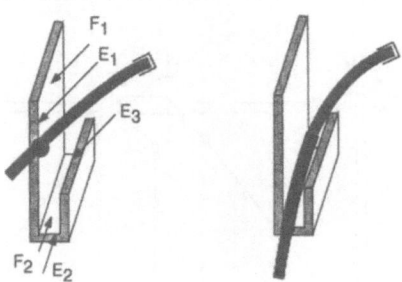

Figure 3: Two states of the insertion process of a hose into an U-shaped guiding groove

Depending on the boundary conditions, there are several possibilities to perform this task. A rather simple and robust procedure is as follows:

- Establish an initial contact between the edge of the hose and edge E_1 of the groove.
- Move the hose downwards without loosening contact until the hose touches either one of the edges E_2 or E_3.
- Establish contact between the hose edge and F_1.
- Move the hose downwards without loosening contact with F_1 until it gets in contact with F_2.

Translating this procedure into a sequence of contact states and state transitions leads to the following:

$$N \rightarrow E/E_1 \rightarrow E/E_1 \wedge (E/E_2 \vee E/E_3)$$
$$\rightarrow E/F_1 \rightarrow E/F_1 \wedge E/F_2$$

Obviously, it would be rather simple to generate a robot program to perform this task if a library of sensor-based, encapsulated routines for the required state transitions was available. Performing assembly tasks in this way requires neither exact knowledge about the mechanical workpiece properties nor a quantitative calculation of the workpiece shape.

---

[2] The occurrence of a second contact is needed as a trigger signal for initiating a new gripper motion. However, the further behavior of the second contacts is not relevant for performing the task.

3. Human Transition Recognition

Based on the ideas introduced above, several experiments are performed in order to gain knowledge about the behaviour of the deformable object and to detect contact state transitions of the workpiece. At first, a list of features is derived from the manual manipulation experiments. Secondly, characteristic measurement curves are generated by observing the features during a state transition. The manual manipulation experiments are verified with a robot and two technical sensor systems in the next section.

3.1 Sensor data features

In the first experiment, a human grips a low elastic workpiece, which belongs to the material class E- defined in [20], for example a pneumatic wire. Those workpieces are deformed elastically when the deforming force is greater than gravity. As obstacles for the manipulation, cubic and pyramidal objects are used similar to the objects shown in Figure 2. Then, each transition of the graph between two stable states with one unstable state in between is examined by a human simulating some sensors, which are in our case visual sensors (a stationary camera with several different viewpoints and a hand camera) and a force/torque sensor. The observed features are one-dimensional values that are directly measured by the sensor or derived from those low-level data. Examples are one single coordinate of the free workpiece endpoint or the angle of the tangent in this point.

3.2 Characteristic functions

By observing manually initiated state transitions, we find that the change of the values of the (one-dimensional) features always can be assigned either to a characteristic feature change description $L(a,b,c,d)$ or to a description $P(a,b,c,d)$ which are both defined in Table 1. $L(a,b,c,d)$ is a useful description when it can be assumed that the curve of feature values is piecewise linear. This is found to be true for slow workpiece motions, for smoothing of the values, and when there is no sudden stress release.

Type	Parameter	Condition
L	$a \; :=$	$\mathrm{sign}(\lim_{t \to t_0+} f(t) - \lim_{t \to t_0-} f(t))$
	$b \; :=$	$\mathrm{sign}(\lim_{t \to t_0-} \Delta f(t))$
	$c \; :=$	$\mathrm{sign}(\lim_{t \to t_0+} \Delta f(t))$
	$d \; :=$	$\mathrm{sign}(\lim_{t \to t_0+} \Delta f(t) - \lim_{t \to t_0-} \Delta f(t))$
P	$a \; :=$	$f(t \leq t_0)$
	$b, c:$	$f(t > t_0) = f(t_0) + b \, \sin(\omega t) e^{-ct}$

Table 1: Definition of the parameters $L(a,b,c,d)$ and $P(a,b,c)$ characterizing the changes in the curves of the feature values

Note that a,b,c,d are defined as elements of the set $\{-1,0,+1\}$. If there is a sudden stress release and oscillations in the workpiece occur, $P(a,b,c)$ can be used to describe the feature value curve. Examples for some characteristic curves are provided in Figure 4.

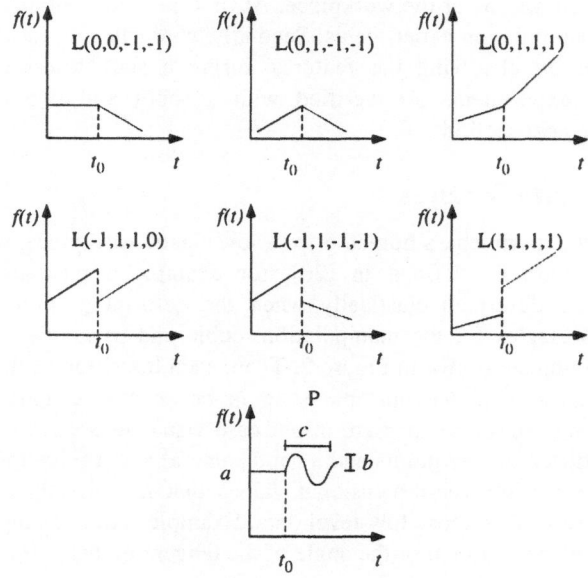

Figure 4: Examples for the observed characteristic changes of features $f(t)$. L(0,0,-1,-1): change after constancy, L(0,1,-1,-1): roof/valley, L(0,1,1,1): slope change, some jumps L($\pm$1,...), and the general look of a curve of type $P(a,b,c)$

3.3 Human observation

One result of the manual experiments concerning the vision sensor is shown in Table 2. In the table, all state transitions that occur between two stable states are classified according to the four visual features curvature $u(t)$, endpoint $p(t)$, tangent angle in the endpoint $g(t)$, angle between endpoints $n(t)$. Due to the simulation of low sample rates, oscillations that would result in a $P(a,b,c)$-curve are not observed.

For the theoretical force/torque analysis we assume an *ideal force/torque sensor*, i.e., a sensor with infinite resolution and without internal noise. Furthermore, we do not consider friction.[3] In a first practical validation, where the tactile sense of a human hand joint is used as a substitute for a wrist-mounted technical force/torque

---

[3] Some additional assumptions concerning the contact state transitions which are of minor importance here are given in [21].

sensor, we find all the classes of feedback curves defined in Figure 4. Similar results as in Table 2 can be expected.

Transition	Characteristic feature change for stationary vision observation			
	$u(t)$	$p(t)$	$g(t)$	$n(t)$
N→V/F	$(0,0,c,d)$	$(0,\pm1,0,\pm1)$	$(0,0,c,d)$	$(0,0,c,d)$
N→E/E	$(0,0,c,d)$	$(0,\pm1,\pm1,\pm1)$	$(0,0,c,\pm1)$	$(0,0,c,\pm1)$
N→E/F	$(0,0,c,d)$	$(0,\pm1,\pm1,\pm1)$	$(0,0,\pm1,\pm1)$	$(0,0,1,\pm1)$
V/F→N	$(0,\pm1,0,\pm1)$	$(0,0,c,d)$	$(0,\pm1,0,\pm1)$	$(0,\pm1,0,\pm1)$
V/F→E/F	$(0,\pm1,0,\pm1)$	$(0,0,c,d)$	$(0,\pm1,0,\pm1)$	$(0,\pm1,0,\pm1)$
V/F→V/E→N	$(\pm1,b,0,\pm1)$	$(\pm1,b,c,d)$	$(\pm1,b,0,\pm1)$	$(\pm1,b,0,\pm1)$
V/F→V/E→V/F	$(0,\pm1,\pm1,\pm1)$	$(0,\pm1,\pm1,\pm1)$	$(0,\pm1,\pm1,\pm1)$	$(0,\pm1,\pm1,\pm1)$
V/F→V/E→E/E	$(0,\pm1,\pm1,\pm1)$	$(0,\pm1,\pm1,\pm1)$	$(0,\pm1,\pm1,\pm1)$	$(0,\pm1,\pm1,\pm1)$
E/E→N	$(0,\pm1,0,\pm1)$	$(0,\pm1,\pm1,\pm1)$	$(0,\pm1,0,\pm1)$	$(0,\pm1,0,\pm1)$
E/E→E/F	$(0,\pm1,0,\pm1)$	$(0,\pm1,0,\pm1)$	$(0,\pm1,0,\pm1)$	$(0,\pm1,0,\pm1)$
E/E→V/E→N	$(\pm1,b,0,\pm1)$	$(\pm1,b,c,\pm1)$	$(\pm1,b,0,\pm1)$	$(\pm1,b,0,\pm1)$
E/E→V/E→V/F	$(0,\pm1,\pm1,\pm1)$	$(0,\pm1,\pm1,\pm1)$	$(0,\pm1,\pm1,\pm1)$	$(0,\pm1,\pm1,\pm1)$
E/E→E/V→N	$(\pm1,b,0,\pm1)$	$(\pm1,b,c,\pm1)$	$(\pm1,b,0,\pm1)$	$(\pm1,b,0,\pm1)$
E/E→E/V→E/E	$(0,b,c,\pm1)$	$(0,b,c,\pm1)$	$(0,b,c,\pm1)$	$(0,b,c,\pm1)$
E/E→E/V→E/F	$(0,b,c,\pm1)$	$(0,b,c,\pm1)$	$(0,b,c,\pm1)$	$(0,b,c,\pm1)$
E/F→N	$(0,\pm1,0,\pm1)$	$(0,\pm1,\pm1,\pm1)$	$(0,\pm1,0,\pm1)$	$(0,\pm1,0,\pm1)$
E/F→V/F	$(0,b,b,0)$	$(0,\pm1,0,\pm1)$	$(0,b,b,0)$	$(0,b,b,0)$
E/F→E/E	$(0,0,\pm1,\pm1)$	$(0,0,\pm1,\pm1)$	$(0,0,\pm1,\pm1)$	$(0,0,\pm1,\pm1)$
E/F→E/→VN	$(\pm1,b,0,\pm1)$	$(\pm1,b,c,\pm1)$	$(\pm1,b,0,\pm1)$	$(\pm1,b,0,\pm1)$
E/F→E/V→E/E	$(0,b,c,d)$	$(0,b,c,d))$	$(0,b,c,d)$	$(0,b,c,d)$
E/F→E/V→E/F	$(0,b,c,d)$	$(0,b,c,d)$	$(0,b,c,d)$	$(0,b,c,d)$

Table 2: Characteristic changes of four visual features for each of the contact state transitions

3.4 Observation with technical sensors

In the case of "real" sensor systems, we again obtain measurement value curves that can be qualitatively characterized by the classes defined in Figure 4. This holds true for all possible state transitions and special measurement values of two sensors: a CCD camera observing the DLO from a stationary viewpoint and a 6 DOF force/torque sensor mounted at the wrist of an industrial robot. The main difference with respect to the curves from the human observation is the presence of noise. Periodic and damped oscillations in the curves coming from released inner stress of the DLO are only observed if the sample rate of the sensor is high enough.

In order to see what is measured during a state transition, we need to analyze for example the initiated transition N→V/F. As similar observations are made for the vision system (see next Section), we only consider the case with a force/torque sensor. Let $F(t)$ and $M(t)$ be the vectors of force and moment measured by the force/torque sensor. Assuming that both force and moment are zero when the DLO is in state N, we can observe the following: When the workpiece touches the obstacle face, it is deformed, causing force $|F|$ and moment $|M|$ rising in a linear manner. Both $|F|$ and $|M|$ increase as long as the gripper motion is continued. The same result is obtained for transitions N→E/E or N→E/F. During the complete motion, F is aligned with the normal of the obstacle face.

Experiments with other transitions show that any state transition is generally accompanied by a distinct change in the course of these functions [23]. Therefore, the task of detecting contact state transitions can be transferred into the problem of detecting these changes in the force and moment signals. The next section introduces our approach for the automatic transition recognition based on the automatic detection of those signal changes.

4. Automatic Transition Recognition

This section deals with the automatic state transition recognition that is used to perform automatically sensor-based robot operations. Again, we use the two different sensors: CCD-camera and force/torque sensor.

4.1 General concept

Since we believe that robust manipulation is possible without geometric reconstruction and without an exact model of the workpiece and the scene, our focus is on measuring features of the workpiece, especially the features stating any deformation in the workpiece. The deformation in the workpiece is detected by the change in several visual and force/torque features, which are extracted from the workpiece shape in the image space or the inner forces and torques resulting in a measureable stress. Since deformation can mean a change in the workpiece state in the context of one state model, changes of the measured features of the workpiece and knowledge about the initial workpiece state are used to derive the current contact state as illustrated in Figure 5. Please note that the coarse obstacle geometry is a-priori given, since it is assumed not to change and therefore can be easily assigned by the programmer or an environment data base.

In our context, sensori-motor primitives transfer a deformable workpiece from a given initial (A) stable contact state to another, desired contact state (E) with guarded robot moves: during the robot motion, the workpiece features are evaluated and used to recognize the current workpiece state. Based on the current state estimation, the robot is moved by a sequence of linear movement commands that aim to transfer the workpiece to a desired final state. Generally speaking, a sensor feature-based, sensor-driven robot control is built. The core of the control are task-independent sensori-motor primitives which here are also referred to as *manipulation skills* [24, 19]. In this work, sensori-motor primitives control the state of the deformable workpiece. In general, these primitives have a controller-like structure as illustrated in Figure 6 and can be combined to task dependent robot operations like such as the threading of a workpiece through a hole.

The next section describes the details of the feature extraction for the two sensors vision and and force torque sensor and a further section shows details of the recognition process.

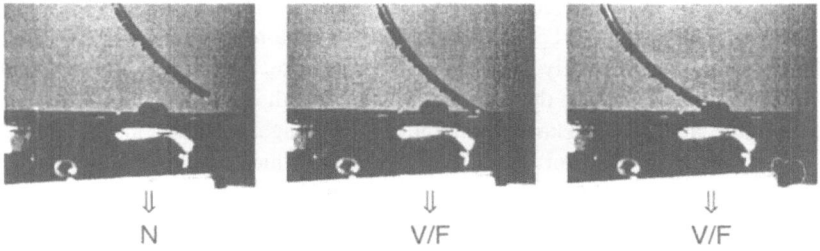

Figure 5: Recognition of workpiece states in an image sequence

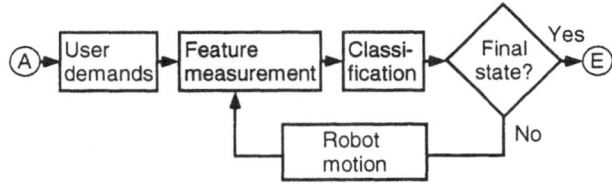

Figure 6: Structure of sensori-motor primitives with initial state (A) and final state (E)

4.2 Feature extraction

This section explains how to extract the features for measuring and recognizing a transition from the sensors. At first, the visual segmentation is discussed. Secondly, the analysis of the force-torque sensor is regarded.

4.2.1 Visual feature extraction

Up to now, many approaches realizing different tasks of a visually guided robotic system are stated in literature [25]. However, there are only few approaches concerning deformable objects, in spite of many important industrial applications [1]. However, it is obvious that vision sensors are well-suited for observing characteristic shape changes of deformable objects. In that way, the information provided by vision sensors is complementary to that of other sensors as for example force/torque sensors. But how can the object be segmented from the image? And which image features can be used for state transition recognition of the object?

In order to get information about the shape of a deformable workpiece, it is necessary to segment the workpiece from the image background. This is commonly known as segmentation. Here, it is assumed that the workpiece is observed from a stationary viewpoint where characteristic shape changes occur in the image.[4] After the segmentation, the deformable workpiece has to be characterized and so some features of the visual data have to be derived. Thus, we need to handle two tasks: workpiece segmentation and feature extraction.

Since there is no need to extract the obstacles from the image and a stationary camera is given, we are using a differential image method to get an idea where the main information about the deformable workpiece is. For that, the current image of the scene including the workpiece is subtracted from a reference image of the scene without workpiece. Currently, we also have some extraction algorithms for the hand camera which work without the differential image method.

Assuming a nearly constant illumination, the image information is furthermore reduced by binarizing the current differential image which results in an image where the workpiece is black and the background completely white or vice versa. The threshold for binarization is determined by taking a value near the mean gray value of the differential image. Even changing illumination is allowed if the pixels not belonging to the workpiece of the differential image are taken as new reference image pixels. After the workpiece image has been segmented, the next step is to generate a representation of the workpiece from which features can be derived which describe the deformation. In the case of linear workpieces, the curve of the projected shape is characteristic for the deformation state. Thus, data from the image of the workpiece represent the workpiece curve. Unfortunately, even thin linear workpieces produce an image with several pixels of extension. In order to get a single curve, the workpiece has to be thinned with thinning operations. But since thinning operations are time consuming iterative operations and tend to be not robust against perturbations in the image, we developed a new algorithm based on a contour following principle. The algorithm searches characteristic image points along the image border of the workpiece. The binarization has not to be made over the whole image, and thus, we save time by performing the binarization locally while searching for the contour. Further details can be found in [26].

Now the workpiece is approximated by points that lie on one borderline of its image. With little effort, it is possible to collect base points of the workpiece that have the same pixel distance. The Manhattan norm is used as distance value. Characteristic base points with different distances are already given by the corner points of the rectangles. Figure 7 shows a workpiece with detected base points. A similar but more complex algorithm has been developed for the hand camera. This algorithm computes also a base point list similar to the one for the stationary camera. Thus, the representation of the workpiece using a base point list is

---

4   Working in the image space allows us to change the camera position without needing a subsequent calibration process.

independent from the used sensor although some particular features derived from the base point list may behave different for the two different vision sensors.

Figure 7: With the stationary camera detected pneumatic wire and found base points (white dots)

Using the base point list as a representation of the workpiece in the image, the features derived from this list give hints to the contact state of the workpiece. These hints come up during the time of robot manipulation. Thus, dynamic features are regarded. Note that the initial and final workpiece state from the graph are given by the operator.

With a list of base points of the two-dimensional image space, three types of characteristic features are considered. These features concern with (1) the start or endpoint of the list, (2) one point within the list, or (3) the whole list of base points. Experiments with our vision system show that features of type (1) and (3) give the most useful information about the workpiece state for a stationary camera as well as for a hand camera.

The following list provides examples for these types of features, which can be efficiently computed:

- The pixel length $l(t)$ of the workpiece. Here, the length is the sum of the lines between the base points in pixels.
- The angle $a(t)$ between the line through the two endpoints and the image axis u.
- The coordinates $p(t)=(u(t),v(t))$ of the endpoint not gripped are used for detecting changes of the contact state of the endpoint. Changes depend on the geometric relation between the workpiece and the workpiece endpoint.
- The tangent angle $g(t)$ in the endpoint indicates a change of the geometric relation between endpoint and obstacle.
- The maximum curvature or the sum of curvatures where the curvature of a base point (u_i,v_i) is approximated by the discretization of the curvature k for regular curves f according to [27]:
- $$k(u_i,v_i) = \frac{\Delta f(u_i,v_i)\Delta^2 f(u_i,v_i) - \Delta^2 f(u_i,v_i)\Delta f(u_i,v_i)}{\sqrt[3]{\Delta f^2(u_i,v_i) + \Delta^2 f(u_i,v_i)}}$$

The first experiments reported in the next section show the applicability of this approach for robust state transition detection, when appropriate features are combined.

Having discussed the visual feature extraction, it now shall be considered how the features for state transition detection may be extracted by means of a (non-ideal) force/torque sensor mounted at the robot wrist.

4.2.2 Force/torque feature extraction

So far, we have discussed the forces caused by the interaction between DLO and obstacle. However, in most practical situations, there is a significant additional load due to gravity. As long as we restrict the gripper trajectory to translational motions, the gravity only causes a constant offset for both the force and the moment. Therefore, it does not have to be considered if absolute thresholds are avoided in the detection algorithm.

Opposite to the handling of rigid workpieces, especially two additional problems must be regarded: First, the force caused by contact with the rigid environment is generally low. This results in the need for a high sensor resolution. However, the resolution is correlated with the measuring range (A high resolution requires a small measuring range and vice versa). Therefore, the measuring range should be as small as possible. Because the force and moment caused by gravity is typically much larger than the contact force, the required measuring range (and, thus, the obtained resolution) is determined by gravity.

Secondly, all contact forces highly depend on both the workpiece and specific situation. Therefore, it is necessary to either have rather precise a priori knowledge about the workpiece and the task, or to base the transition detection on the qualitative coarse of force and moment while avoiding absolute thresholds. In order to meet the fundamental requirements given for manipulation skills, the second approach should be preferred.

As the first step of the signal processing, some filtering is required for noise reduction. In our experiments, we succeeded in using a moving average low pass filter which can be implemented very efficiently for our purpose. Below, we generally refer to F and M as filtered force and moment signals.

For evaluating the 6D force/moment vector, there are several possibilities. We succeeded in evaluating the absolute values of force and moment or in evaluating the six single signals independently. The directions of the vectors do not have to be considered here because it does not promise to be advantageous and the effort for a vectorial evaluation is rather high. In the following discussion of the detection algorithms, we always refer to a general function f, which is sampled. This may be either some visual features or a single component of F or M, or the absolute values of these vectors.

4.3 State transition recognition: detection of characteristic functions

Now, the following two questions are investigated: how can state transitions be detected with a sensor system (vision and force/torque sensor) and what is necessary to perform the state transitions reliably and robustly with a robot system?

By observing state transitions executed by the robot and the corresponding feature values processed and analyzed by our sensor systems, we find that the change in the values of the (one-dimensional) features follows the same pattern as for the manual experiments. As a consequence, we can use them as well here. This means especially that we have to provide at least two state transition detection algorithms, at least one for each curve type L and P.

Fortunately, through several different experiments for each the vision and force/torque sensor, we found that almost the same algorithms can be applied on both sensor systems. Only some parameters of the algorithms change depending on the sensor. The algorithm parameters are also dependent on the current environment. In the following, the two algorithms, which are still in the test phase, are described. further details can be found in [23].

4.4 Sensor signal processing

After the analytical discussion of the state transitions in the previous section, it shall be considered how they may be detected by means of a (non-ideal) vision sensor (stationary camera or robot hand-mounted camera) or a (non-ideal) force/torque sensor (FTS) mounted at the robot wrist. Before presenting the evaluation algorithms, some general aspects are discussed.

The detection of the $L(a,b,c,d)$ **transitions** can be traced back to the problem of detecting (more less abrupt) changes in the slope of the feature curve f provided from a certain sensor. For the moment, we assume a function with always positive slope and the slope being higher for $t > t_0$ than for $t < t_0$.

The detection algorithm is based the fact that the slope $f'(t)$ of function $f(t)$ changes rather abruptly for $t = t_0$. For the ideal case of two linear segments, $f'(t)$ is a saltus function. Since this is not the case in practical applications and there is generally some noise remaining after low-pass filtering, we use the following function instead of $f'(t)$. Let Δt be the sampling period, and be $f(t) = f(i\Delta t) = f_i$ the function value sampled in the current time step i, $m_{\mathrm{MSL},i}$ is the slope of the mean straight line (MSL) computed from the $k+1$ samples $i-k$, $i-k+1,...,$ i by a standard algorithm. Based on m_{MSL}, we define

$$\Delta m_{\mathrm{Ratio},i} = \frac{m_{\mathrm{MSL},i} - m_{\mathrm{MSL},i-k}}{m_{\mathrm{MSL},i-k}}$$

as the relative slope change of two MSLs touching at i_0.

To clarify the characteristics of $\Delta m_{\mathrm{Ratio}}$, Figure 9 shows the values f_i of a discrete function vector f with slopes $m_{\mathrm{left}} = 1$ and $m_{\mathrm{right}} = 4$ together with $\Delta m_{\mathrm{Ratio},i}$. The mean straight lines are computed with $k+1 = 21$ samples. At the beginning, $\Delta m_{\mathrm{Ratio}}$ is 0, i.e., both slopes in the nominator of $\Delta m_{\mathrm{Ratio}}$ have the same value. As i is ap-

---

5   In both cases the gripper motion is restricted to linear motions because rotation generally influence the measured moment, and thus, may lead to the erroneous detection of transitions.

proaching i_0+k, Δm_{Ratio} increases. For $i = i_0+k$, it reaches an extremum which is equal to the relative slope ratio of m_{left} and m_{right}. For $i > i_0+k$, the function approaches zero again with $\Delta m_{\text{Ratio}} = 0$ for $i > i_0+2k$. Using this function, the task of detecting a slope change in f is transformed into the task of detecting an extreme value of vector Δm_{Ratio}.[6]

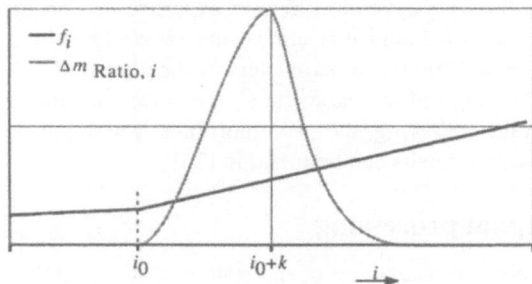

Figure 8: Discrete function f_i consisting of two linear segments and relative difference Δm_{Ratio} of the mean straight line

However, the course of Δm_{Ratio} has typically much more extrema than there are state transitions to be detected. Therefore, we use the following algorithm with some additional criteria. The vectors m_{MSL} and Δm_{Ratio} are defined as given above, $b_{\text{MSL}, i}$ is the intercept of the MSL in sample i, defining the MSL together with $m_{\text{MSL}, i}$.

```
function IsTransition(i: integer): boolean;
    const σmax, MaxDevmax, εSearch
           ΔmRatio, min, Δmmin;
        begin
        fi := ReadSensorData;
        for j := 0 to k do
Devj := |fi-k+j - (mMSL, i * j + bMSL, i)|;
        σ := StdMeanDev(Dev);
        MaxDev := max(Dev);
        IsValidi :=      σ ≤ σmax
           and     MaxDev ≤ σ*MaxDevmax;
        ii := i - εSearch;
    IsTransition:=       IsValidii
        and             IsValidii-k
        and     IsExtreme(ΔmRatio, ii, εSearch)
        and       | ΔmRatio, ii | ≥ ΔmRatio, min
           and  mMSL, ii − mMSL, ii-k ≥ Δmmin;
        end;
```

---

[6] On the one hand, the sharpness of the extreme value is increased with k decreasing. On the other hand, the susceptibility to distortions is increased, too.

After sampling a new data value f_i from the sensor, we compute the MSL given by the samples i-k, ... i. If the following two conditions are fulfilled, we suppose the sampled data to be valid ($IsValid_i$: = **true**):

- the standard mean deviation σ between the data values and the MSL is not greater than a threshold σ_{max}, and
- the maximum deviation $MaxDev$ from the MSL is not greater than the $MaxDev_{max}$-fold of σ.

Otherwise, the data are considered to be corrupted by noise or other distortions ($IsValid_i$:= **false**).

After performing this check, $\Delta m_{Ratio, i}$ is computed as described above. Please note that the algorithm is based on a search for extreme values of Δm_{Ratio}. Therefore, a state transition in sample i_0 can not be detected immediately, but only in sample i_0+k+ε_{Search}, with $2\varepsilon_{Search}$ being the width of the window used for the extremum search. (In the opposite to the handling of rigid materials, a delayed detection of state transitions is less critical for deformable objects.) Therefore, the transition detection is performed for sample $ii = i$-ε_{Search} in step i.

For a state transition, the value sampled in step ii must fulfill the following conditions:

- For both MSLs used for computing $\Delta m_{Ratio, i}$, the condition $IsValid$ must be fulfilled. That is, the regarded area of the data array must not be severely disturbed.
- The function vector Δm_{Ratio} must suppose an extremum at position ii as discussed above.
- Both the absolute value of $\Delta m_{Ratio, ii}$ (that is, the relative slope difference of the MSLs) and the absolute slope difference Δm_{ii} must be greater or equal than given threshold values $\Delta m_{Ratio, min}$ and Δm_{min}, respectively.

Figure 9 clarifies these conditions for a N $\rightarrow$ V/F transition, using the absolute value of the moment vector M as measuring signal.

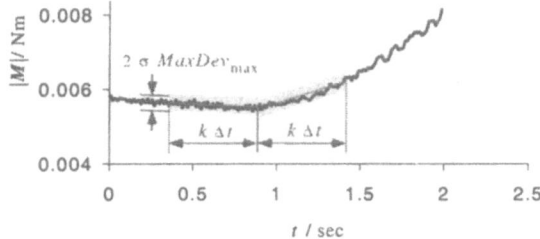

Figure 9: Absolute value of moment M for a transition N $\rightarrow$ E/E

In the formulation given above, the algorithm considers only one case of all possible slope changes. According to some above assumptions, the slope of the considered function may be either positive, negative, or zero for both $t < t_0$ and

$t > t_0$. Therefore, the algorithm must be extended. The algorithm given above can be adopted to the detection of all kinds of transitions by considering the signs of Δm_{Ratio} and Δm.

For detecting **oscillations P(a,b,c)**, we have to remind the fact that none of its characteristics, i.e., frequency, amplitude and damping, are generally known in advance. Thus, an algorithm is required for the robust on-line detection of arbitrary (sinusoidal) oscillations.

Our detection algorithm is based on the following assumptions:

1. though the oscillation period T is unknown, we can give a lower boundary T_{min} with $T \geq T_{min}$ (and possibly also an upper boundary T_{max}).
2. though both the (initial) amplitude A_0 and damping are unknown, we can give a lower amplitude boundary A_{min} and a number n_{min} with $A \geq A_{min}$ for at least n_{min} consecutive oscillation extrema.

For the detection, we perform an online search for extreme values in the force or moment function. An oscillation is detected when there are n_{min} consecutive extrema, each having a temporal distance of at least T_{min} and an elongational distance of at least A_{min} to the preceding one, as shown in Figure 10 for $n_{min}=3$. Though both the oscillation amplitude and the period depend on the workpiece and the situation, the algorithm proved to work reliably for a large variety of workpieces and situations without changing A_{min} and T_{min}.

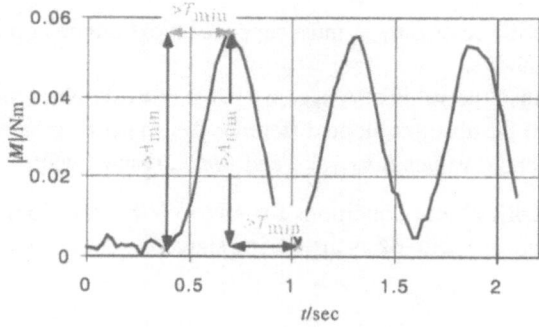

Figure 10: Amount of moment vector M for a V/F→V/E→N sequence. For clarity reasons, the static load due to gravity is not shown

5. Experimental Results

In this section, results of the sensor-based robot manipulation of deformable linear objects using the state transition recognition approach are presented. Both sensor systems are used: machine vision and force/torque measurement.

5.1 Vision-based workpiece manipulation

For the robot manipulation experiments, a Kuka KR15 robot is used. The robot controller executes motion commands sent from a Linux-PC with 350 MHz Intel Pentium II Processor. As stationary camera, a standard video CCD-Camera (Hitachi KP M3) is used, and as hand camera a Teli CS 3710 C is used. The data are sent by a standard frame grabber (Eltec PC-EYE I) to the vision processing computer which is a Linux-PC with either a 133 MHz or a 350 MHz Intel Pentium II Processor. As force torque sensor an experimental sensor of the "Deutsche Luft- und. Raumfahrtgesellschaft" (DLR) is used. The gripped workpiece is a pneumatic polyurethane wire with the outer diameter of 6 mm and a length of 300 mm. The obstacle is a car door frame which is mounted in a horizontal lying position. Details of the software system and its structure can be found in [28].

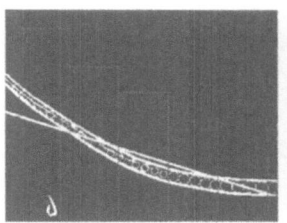 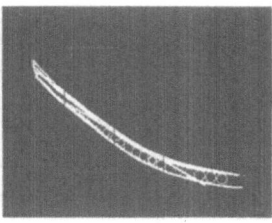 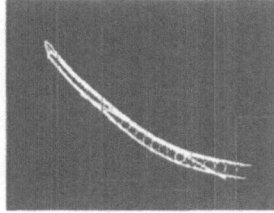

Figure 11: A sequence of images where a pneumatic wire is segmented and tracked with the hand camera

The segmentation and feature analysis algorithms were tested in various experiments and work well under both daylight and artificial lighting conditions. Measurements yield a mean execution time of 1.1ms on an Intel Pentium machine with 133 MHz processor when a workpiece is detected. For common situations and their well-known problems, usually a small set of parameters has to be adapted what can be done from an expert in a few minutes. Figure 11 shows a sequence of images where a pneumatic wire is segmented and tracked with the hand camera by using a gray value difference detection with the Sobel operator.

Since the force/torque sensor produces very similar feature data compared to both vision sensors, subsequently only experiments involved in the detection of a state transition with our force/torque sensor will be reported. Furthermore, the algorithms described here have been investigated for both isolated state transitions and an assembly task.

5.2 Force/torque-based workpiece manipulation

As robot manipulator, a KUKA KR 15 industrial robot with PC-based KR C1 robot controller has been used. As DLOs, we used a Polyurethane pneumatic wire with 6 mm outer and 4 mm inner diameter, a spring steal wire of 1 mm diameter, and a commercial spring steal ruler of length 0.5 m and (18×0.5) mm^2 cross section. Due to the bending rigidity of the objects, the ruler shows the highest contact forces, while they are smallest for the pneumatic hose. Accordingly, the difficulty of detecting state transitions increases in this order, too.

As an example, Figure 9 shows the absolute value of moment M for a sequence N → E/F →E/E with motion direction MD_1 according to Figure 3 for the steal ruler and the hose. Though the contact force is considerably higher for the steal ruler, the transitions were detected correctly for both materials with adopting only the parameter Δm_{min}. Compared to the steal ruler, the detection of the N → E/F transition is critical for the hose because the force caused by the E/F-contact is very small. In such cases, the reliability can be considerably improved if the time of the state transition is approximately known. If this condition is fulfilled, erroneous detections of transitions can be avoided by starting the detection algorithm only some time before the detection. Especially when starting and stopping gripper motions, the detection of state transitions should not be active.

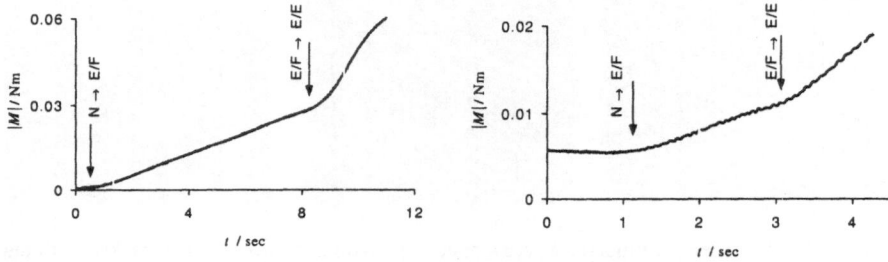

Figure 12: Transition sequence N → E/F →E/E for steal ruler (left) and pneumatic hose (right), using the following parameters: Sampling period $\Delta t = 10$ ms, $k = 80$ samples, $\Delta m_{min} = 1*10^{-5}$Nm (ruler) respectively $\Delta m_{min} = 2*10^{-5}$Nm (hose), $\Delta m_{Ratio, min} = 0.5$, $\sigma_{max} = 4*10-5$ Nm, $\varepsilon_{Search} = 15$ samples

Compared to the moment shown here, the force signal is often found to be rather noisy and could hardly be evaluated. This is due to the fact that the forces to be measured are rather small. However, because the distance between the gripper and the contact point is rather long, even these small forces cause significant moments. Especially all disturbances caused by the (rather high) inertia in the gripper impact on the force much more than on the moment.

In addition to the investigation of isolated transitions, we performed the task of inserting a pneumatic hose into a guiding groove as described in [21] and used the algorithm presented here for detecting the contact state transitions. The algorithm proved to be reliable in this task. However, we find that the detection is less critical for the transitions with the contact point being rather close to the gripper than for transitions with the contact point (point of force application) being far from the gripper. This is caused by the fact that the bending rigidity of the workpiece increases with decreasing length between the gripper and the contact point. Especially for low contact forces, some problems are caused by friction which may be of larger influence than the change of contact force. Thus, the friction coefficient between the workpiece and the obstacle should be as small as possible.

6. Conclusions and Future Work

In the assembly of rigid workpieces, the consideration of contact states and state transitions has been proven to be a suited method for creating robust manipulation routines for multiple purposes. We expect that this is valid for deformable workpieces as well. In this paper, we investigate contact states and state transitions of deformable linear objects in a rigid environment. The application to a typical industrial problem shows that this principle can be easily used for describing assembly tasks.

In this chapter, a new and systematic approach to the machine vision-based and force/torque-based robot manipulation of deformable linear workpieces is proposed. This powerful approach reduces the computational needs by using a simple state-oriented model of the workpiece.

The states describe the relation between the workpiece and an obstacle, and they are derived from the sensor data and the features delivered by two different sensor systems (vision and FTS). Using the vision sensors, the workpiece image is segmented from a standard video frame using a new and fast segmentation algorithm. With the FTS, forces and moments can be measured directly. Finally, for both sensor systems, the workpiece features are computed in order to recognize state transitions during the manipulation of the workpiece by a robot.

Experimental results prove the applicability of our approach. Characteristic feature changes are derived from manual manipulation and observation experiments. They are used for implementation and optimization of the sensor processing and the workpiece state classification. Two state transitions that are recognized reliably with our system are presented.

Because there are only two main types of characteristic functions (oscillations and slope changes), two detection algorithms were developed for detecting the transitions. Both of them where found to be reliable in an experimental investigation. However, the robustness of the detection depends on several physical and technical parameters as for example the workpiece rigidity, friction coefficients, and the sensor resolution.

For future work, we need to further investigate on which parameters can be computed automatically and how can their number be minimized. We have to answer the question which sensor system is useful for which transition and under which conditions. This will also include a comparison of the results of our research for every sensor system. As further step, we will concatenate the implemented vision-based robot primitives for deformable linear objects in order to execute complex manipulation tasks.

The detection algorithms presented here contain some "hard" threshold values that need to be set according to the specific task. Additionally, the detection may fail in the presence of negligible uncertainties. Therefore, we propose that a fuzzification of the algorithms will improve both the generality and reliability of the state transition recognition. For improving the reliability when contact forces are small, we will investigate the usage of force/torque sensors with very high resolution. The next major step is to construct encapsulated routines for performing manipulation tasks, including both gripper motions and transition detection.

7. References

[1]   Byun J.-E., Nagata T.: "Determining the 3-D pose of a flexible object by stereo matching of curvature representations". In: Pattern Recognition: The Journal of the Pattern Recognition Society, vol. 29, no. 8, pp. 1297-1308, 1996.

[2]   Nakashima M., et al.: "Application of semi-automatic robot technology on hot-line maintenance work". In: IEEE Int. Conf. on Robotics and Automation, pp. 843-850, 1995.

[3]   Hirai, S., Wakamatsu, H., and Iwata, K.: "Modeling of deformable thin parts for their manipulation". In: Proc. 1994 Int. Conf. on Robotics and Automation (ICRA'94), vol. 4, pp. 2955-2960, San Diego, USA, May 1994.

[4]   Wakamatsu, H., Hirai, S., and Iwata, K.: "Modeling of linear objects considering bend, twist and extensional deformations". In: Proc. 1995 Int. Conf. on Robotics and Automation (ICRA'95), vol. 1, pp. 433-438, Nagoya, Japan, May 1995.

[5]   Wakamatsu H., et al.: "Dynamic analysis of rodlike object deformation towards their dynamic manipulation", In: Proc. 1997 IEEE/RSJ Int. Conf. on Intelligent Robots and Systems (IROS'97), pp. 196ff, Grenoble, France, September 1997.

[6]   Remde A., Henrich D.: "Direct and Inverse Simulation of Deformable Linear Objects". Chapter in this book.

[7]   Inoue H., Inaba M.: "Hand-eye coordination in rope handling". In: Proc. of the First Int. Symp. On Robotics Research, pp. 163-174, New Hampshire, USA, September 1983.

[8]   Zheng Y. F., Pei R. Chen C.: "Strategies for automatic assembly of deformable objects". In: Proc. 1991 Int. Conf. on Robotics and Automation (ICRA'91), vol. 3, pp. 2598-2630, Sacramento, USA, April 1991.

[9]   Nakagaki H., et al: "Study of insertion task of a flexible wire into a hole by using visual tracking observed by stereo vision". In: Proc. 1996 Int. Conf. on Robotics and Automation, vol. 4, pp 3209-3214, Minneapolis, MN, USA, April 22-28, 1996.

[10]   Kraus W., McCarragher B. J.: "Case studies in the manipulation of flexible parts using a hybrid position/force approach". In: Proc. 1997 Int. Conf. on Robotics and Automation (ICRAí97), vol. 1, pp. 367-372, Albuquerque, USA, April 1997.

[11]   Cheng C., Zheng Y.F.: "Deformation identification and estimation of one-dimensional objects by using vision sensors". In: Proc. of the 1991 IEEE Int. Conf. on Robotics & Automation, Sacramento, California, USA, April 1991.

[12]   Nakagaki H., Kitagaki K., Ogasawara T., Tsukune H.: "Study of deformation and insertion tasks of a flexible wire". In: Proc. of IEEE Int. Conf. on Robotics and Automation (ICRA'97), pp. 2397-2402, Albuquerque, USA, April 1997.

[13]   Smith P.W.: "Image-based manipulation planning for non-rigid objects". In: Proc. 1998 Int. Conf. on Robotics and Automation, vol. 4, pp 3540-3545, Leuven, Belgium, May 1998.

[14]   Hasegawa T., Suehiro T., Takase K.: "A model-based manipulation system with skill-based execution". In: IEEE Transactions on Robotics and Automation, vol. 8, no. 5, pp. 535-544, Oct. 1992.

[15]   Onda H., Hirukawa H., Takase K.: "Assembly motion teaching system using position/force simulator – Extracting a sequence of contact state transition". In: IEEE/RSJ International Conference on Intelligent Robots and Systems (IROS'95), vol. 1, pp. 9-16, 1995.

[16]   Crowell R. H., Fox R. H.: "Introduction to knot theory", Springer New York Heidelberg, 1977, ISBN 0-387-90272-4, 3-540-90272-4.

[17]   Suehiro T., Takase K.: "Representation and control of motion in contact and its application to assembly tasks". In: Proc. 5th Int. Symp. on Robotics Research, 1989.

[18]   Higashijima K., Onda H., Ogasawara T.: "Planning for wire obstacles avoidance using ultrasonic sensors". In: IEEE/RSJ Int. Conf. on Intelligent Robots and Systems (IROS'98), Victoria, October 1998.

[19]   Morrow J. D., Khosla P. K.: "Manipulation task primitives for composing robot skills". In: IEEE Int. Conf. on Robotics and Automation (ICRAí97), pp. 3354-3359, Albuquerque, USA, April 1997.

[20]   Henrich D., Ogasawara T., Wörn H. "Manipulating deformable linear objects – Contact states and point contacts". In: 1999 IEEE Int. Symp. on Assembly and Task Planning (ISATP'99), Porto, Portugal, July 21-24, 1999.

[21]   Remde A., Henrich D., and Wörn H.: "Manipulating deformable linear objects: Contact state transitions and transition conditions". In: 1999 IEEE/RSJ Int. Conf. on Intelligent Robots and Systems (IROS'99), Kyongju, Korea, October 1999.

134

[22]   Morris G. H., Haynes L. S.: "Robotic assembly by constraints". In: Proc. 1987 IEEE Int. Conf. on Robotics and Automation (ICRA'87), pp. 1507-1515, 1987.

[23]   Remde, A., Henrich, D., and Wörn H.: "Manipulating deformable linear objects - Force based detection of contact state transitions –". Submitted to: 2000 IEEE International Conference on Robotics and Automation (ICRA 2000), San Francisco, CA, USA, April 2000.

[24]   Morrow J.D., Khosla P.: "Sensorimotor primitives for robotic assembly skills". In: Proc. 1995 IEEE Int. Conf. on Robotics and Automation, pp. 1894-1899, Nagoya, Japan, May 1995.

[25]   Hutchinson S., Hager G.D., and Corke P.I.: "A tutorial on visual servo control". In: IEEE Trans. on Robotics and Automation, vol. 12, no. 5, October 1996.

[26]   Abegg, F., Henrich, D., and Wörn H.: Manipulating deformable linear objects: Vision-based recognition of contact state transitions". In: Proc. 9th Int. Conf. on Advanced Robotics (ICAR'99), Tokyo, Japan, Oct. 25-27, 1999.

[27]   Gray A.: "Modern differential geometry of curves and surfaces", CRC Press, 1994.

[28]   Henrich D. and Abegg F. and Wurll Ch. and Wörn: "A parallel control architecture for industrial robot cells". In: Proc. 4th International Symposium on Distributed Autonomous Robotic Systems, pp. 77-86, Karlsruhe, May 1998.

Section 3.4

Automated Sewing System and Unfolding Fabric

E. Ono

Abstract. Fabric is flexible and soft and its shape readily and markedly changes three-dimensionally. It is a limp material with many nonlinear, anisotropic and hysteresial properties. These facts make it difficult to handle and its manipulation very different from handling rigid object. Many researchers designed devices to pick up fabric as parts of machines for picking or separating, not a multi-purpose handling device like a human hand. In this paper, automated sewing system and devices developed under a MITI project as well as robotic fabric handling are introduced.

1. Fabric Characteristics

Handling characteristics of fabric are mainly flexibility, softness. Fabric shape readily and markedly changes three-dimensionally. Those facts make it difficult to handle and its manipulation very different from handling rigid object. All fabric involves twisted fiber. Whose shape changes not only due to elongation, bending, and twisting, i.e., fiber and yarn, but also due to structure determined by weaving or knitting. Plain woven fabric that does not stretch, essentially, when pulled along either of its perpendicular woven directions, is markedly deformed if pulled obliquely. Therefore, fabric can cover and wrap spherical surface, but paper cannot. Because paper changes its shape only owing to material characteristics, not structure. Fabric properties of extension, bending, resilience and compression are nonlinear, anisotropic and hysteresial in general [Figure 1]. Environmental condition (temperature and humidity) affects the extension and contraction of fabric. Furthermore, fabric deformability also relates with its size, shape and weight. All those properties make shape prediction of fabric rather difficult.

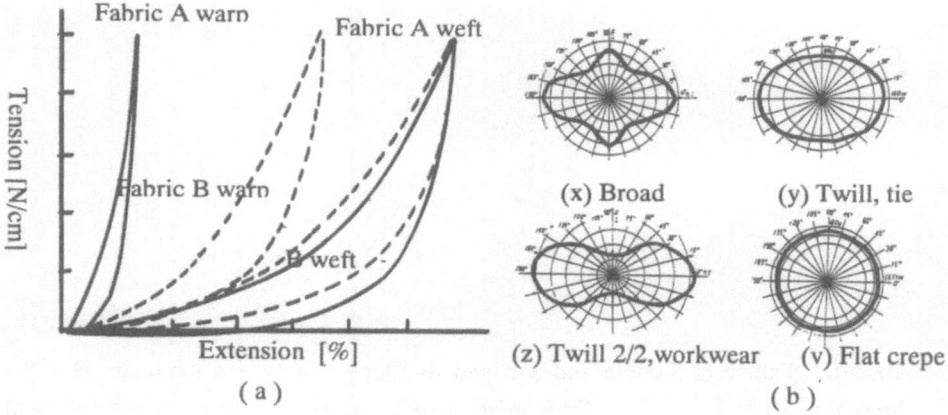

Figure1: (a) Relation between extension and tension. (b) Directional properties of bending rigidity of various woven fabric (x,y,z,v).

2. Automated Sewing System

2.1. Introduction

Automated sewing system was one of large-scale projects of the Agency of Industrial Science and Technology between 1982 and 1991. Three national institutes and twenty-eight companies were involved. Those researched the automated sewing system, which efficiently made apparels in small lots and wide variety. The research target is to reduce production time cycle to a half while under conventional production time cycle, apparels (tops, bottoms, dress, sportswear and night wear) were made at the same time on the same production line. Therefore, the system itself, preparatory sewing technology, sewing and assembly technology, fabric handling technology, and system management and control technology were the subjects under research and development in cooperation with related participant organizations. From fiscal 1988, research and development of a test plant was being carried out. The test plant consists of 4 sub-systems. They are high-speed laser cutting system, flexible sewing system, high-technology assembly system and 3-D flexible press system. Tasks were done from intelligent cloth inspecting before cutting to automatically pressing shape and ladies' blazer was made up. [Figure 2]

2.2. Fabric Handling

In fabric handling technology, reliable separation method from laminated sheets is difficult and so many methods have been invented. [1-2][Figure 3] Under the automated sewing system project, elemental technologies were researched such as devices for picking up and placement, robotic hands and 7 degrees unit arms. [Figure 4] Some of them were used for transfer and placement of fabric in the test plant. [Figure 5]

Author investigated a robotic hand. Most of conventional methods are available on two-dimensional handling tasks, which drag or transfer a plane fabric part not as to change its shape on a workstation. Many research results show how robotic fabric handling is difficult. [3-12] However, a robotic hand is available to manipulate a wide range of materials & to carry out varied tasks. The hand investigated [Figure 4(e)] can sense thickness of fabric and confirm secure separation. The robotic hand has multi-purpose, including separating one piece of fabric from a laminated sheets, picking it up, transferring and placement on 3 dimensional surface not only flat horizontal surface. [13-17] (In next section some pictures are shown about separating, picking, transferring and placement at Figure 6.)

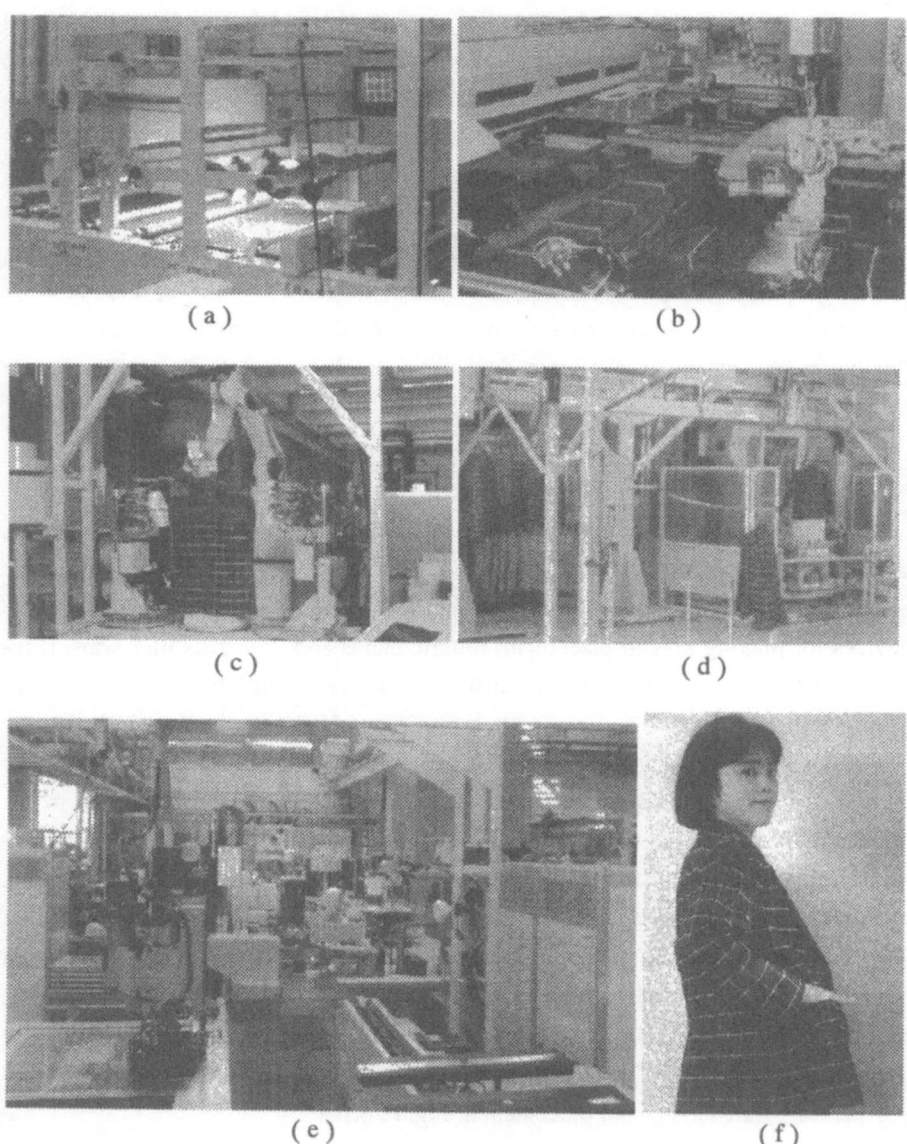

Figure 2 : The test plant of automated sewing system, which was made by connecting some of newly developed technique and machines. (a) A part of Intelligent cloth inspector. (b) Multi-functional sewing station. (c) 3-dimensional seamer system. (d) Flexible press and auto hanger system. (e) Overview of the plant. (f) Lady's blazer made by the test plant, using high-efficient pattern.

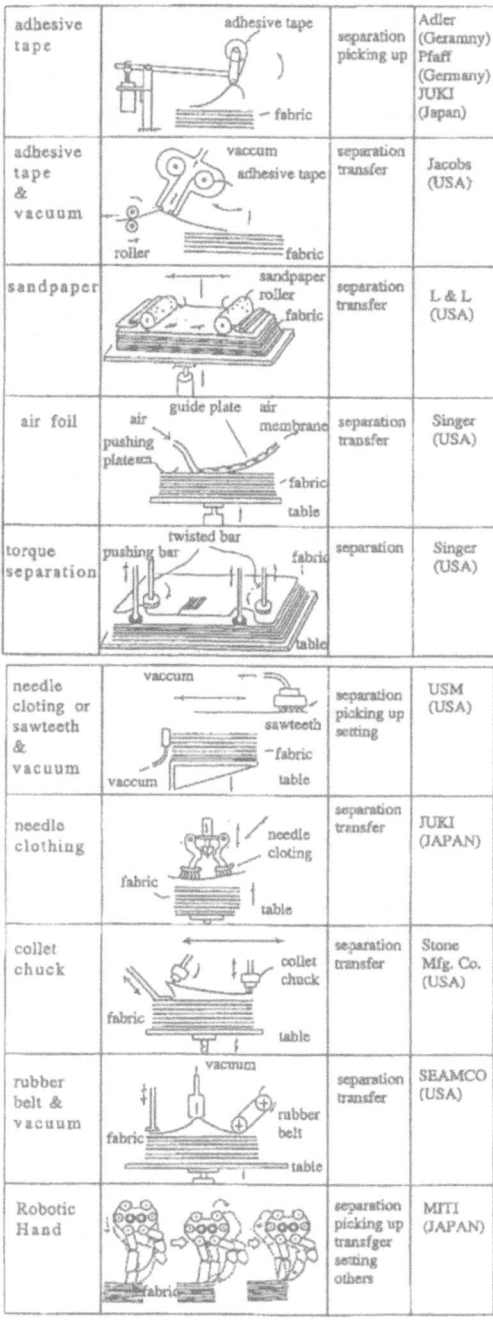

adhesive tape	adhesive tape — fabric	separation picking up	Adler (Geramny) Pfaff (Germany) JUKI (Japan)
adhesive tape & vacuum	vaccum adhesive tape roller — fabric	separation transfer	Jacobs (USA)
sandpaper	sandpaper roller fabric	separation transfer	L & L (USA)
air foil	guide plate air air membrane pushing plate — fabric — table	separation transfer	Singer (USA)
torque separation	twisted bar pushing bar fabric table	separation	Singer (USA)
needle cloting or sawteeth & vacuum	vaccum sawteeth — fabric vaccum table	separation picking up setting	USM (USA)
needle clothing	needle cloting fabric table	separation transfer	JUKI (JAPAN)
collet chuck	collet chuck fabric table	separation transfer	Stone Mfg. Co. (USA)
rubber belt & vacuum	vacuum rubber belt fabric table	separation transfer	SEAMCO (USA)
Robotic Hand	fabric	separation picking up transfger setting others	MITI (JAPAN)

Figure 3: Various separation ideas.

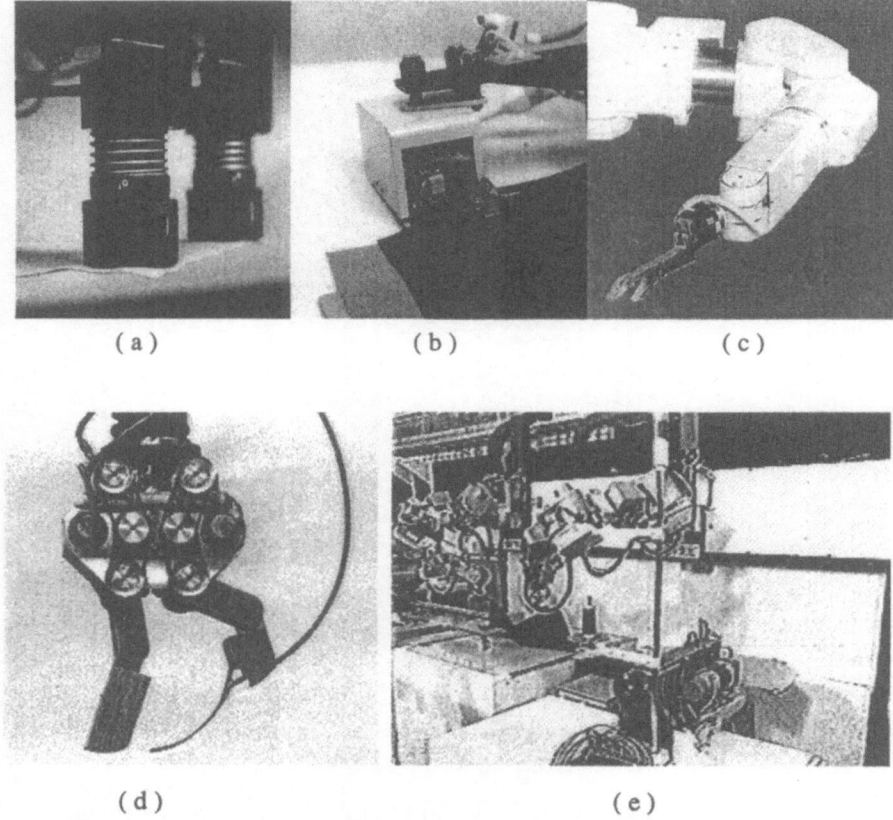

Figure 4: (a) Vertical-holding device: Holds from the upper surface of fabric using needles. (b) Fabric-manipulation device: Holds and manipulates 2 pieces of fabric during sewing. (c) Fabric gripper: pick up a piece of fabric. (d) Robotic hand: Has multi-functions. (e) Module handler: A handler is constructed with modules. Two handlers operate in cooperation with each other.

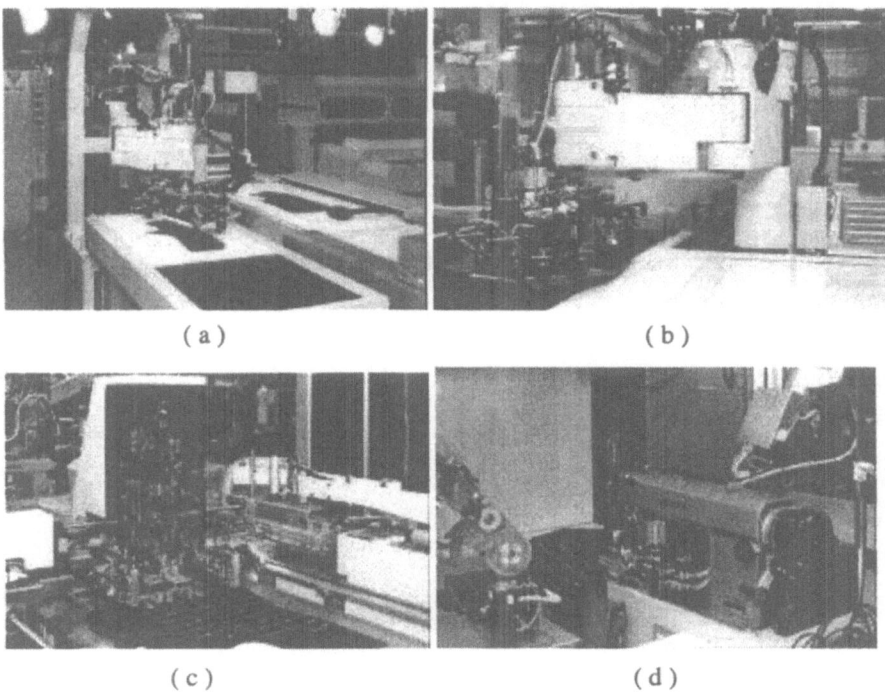

(a)   (b)

(c)   (d)

Figure 5: Several parts of handling sections in test plant. (a) Machine for mating interlining cloth with part. It picks up cut interlining cloth on the conveyor and mates it with part on another conveyor to convey it to the next machine. (b) Automated closing system. It transports front and back bodies to join their sides. (c) Seam opening and pressing system. It conducts seam opening and 3-dimensionaly processes the surrounding of seam sections. (d) Multi-handle cooperative controlled assembly system. It joins the left and right backs with two handles driven in a concerted motion.

We are currently studying flexible material handling that adds a visual sensor to this hand and uses visual and tactile sense, targeting a more flexible system. Latter part of this paper discusses planar unfolding of overlapped fabric. [18]

3. Basic Approach to Handling Deformable Objects Such As Fabric

3.1. Basic Approach

We observed a robot in an experiment, in which it separated a sheet of fabric from a stack and re-stacked it on a table by hooking the sheet on a wire frame and spreading it just before stacking because the robot was single-handed.[Figure 6] We thus realized that handling of flexible objects by robots must properly evaluate the following points:

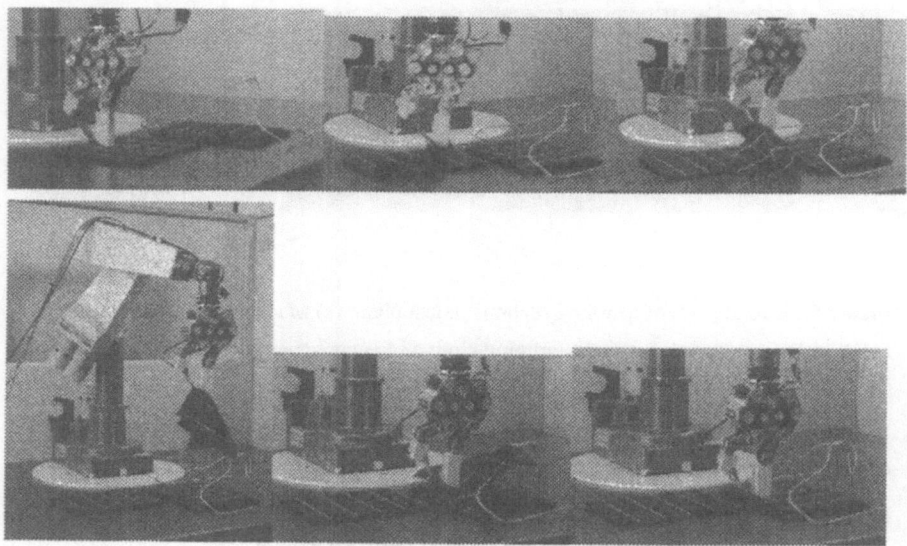

Figure 6: Separating, picking up, Transfer and Placement of a piece of fabric.

(1) To conduct cooperative sensing by visual and tactile sensors.

Visual and tactile sensors must operate complementarily to obtain visual and tactile information for handling flexible objects efficiently due to the ease with which objects change form. More specifically, versatility is increased by incorporating individual visual and tactile sense information or by efficiently using visual and tactile information interactively or simultaneously to optimize information use.

Grasping fabric, for example, is more efficient through seeing than groping alone. More effective ways must be found for grasping after touching because tactile factors such as softness and smoothness and accurate information on weight and center of gravity cannot be obtained purely visually. Auxiliary motion, such as changing the attitude and position of fabric, also enables visual information to be obtained more easily, as does changing the position and attitude of the visual sensor to select tactile information more easily. In handling flexible objects, sensor-based control and motion planning will play a larger role than conventional techniques based on models; i.e., cooperative sensing will be more important in making motion more flexible.

(2) To achieve two objectives in one motion is important.

Correcting grasping in handling flexible objects is done more often than in handling rigid objects, so deformable objects are handled using motions with more than one objective or involve the fewest motions possible to change grasp. The robot we developed turns over and grasps a piece of fabric in a single motion of separation. One may simplify this by grasping the sweater near the neck with one hand and putting the other straight into its sleeve. Combined motions like this can simplify tasks by reducing the number of different movements or changes of grip required, and can therefore greatly aid a robot system.

(3) It is important to determine where to grasp the object by anticipating changes in object shape.

A square fabric grasped at the center on one side and picked up will assume the shape of an inverted V or M with its center at the point grasped. [Figure 7] The shape taken depends on how it is picked up and grasped, the fabric's features, and how the fabric is affected by environmental factors such as humidity. Grasping the fabric elsewhere results in an asymmetric inverse V shape in most cases, which is why the act of grasping must be fully considered.

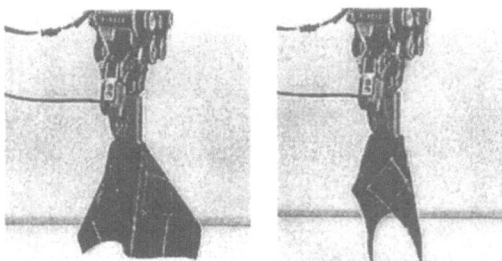

Figure 7 The shape of fabric is not always identical even if the behavior of picking it up is just the same.

(4) Handling must be planned to include trial and error.

Fabric deforms under gravity, and exact shapes are difficult to predict. 3-D deformation is significantly influenced by friction and environment (e.g., humidity), requiring trial-and-error handling. A robot may fail to separate a laminated fabric sheet at first, but succeeds the second time repeating the same motion because the lamination fabric subtly changes its condition in the first trial.

(5) Time-dependent sensor information must be input immediately.

The shape of a flexible object may markedly change due to initial conditions and handling, e.g., from an M to an inverse V, the change depending on the transfer pattern, and such deformation must be sensed as soon as possible to ensure flexibility.

3.2. Unfolding of Overlapped Fabric

3.2.1 Planar unfolding

Planar unfolding from a horizontal plane may be done many ways:

(1) Picking up an edge and shaking and spreading fabric to generate transverse waves [in the way a sheet is spread].

(2) Grasping multiple edges, stretching downward by gravity, and spreading by changing grasp.

(3) Spreading fabric via friction between the fabric and table surface created during sliding.

(4) Holding down one edge on the horizontal plane and slipping and spreading fabric and pulling it outward from where it is held down.

(5) Grasping multiple points and spreading fabric while pulling in the direction grasped for more separation.

(6) Grasping an edge, blowing fluid against the fabric or immersing it, and handling it based on how it spreads.

(7) Grasping an overlapped edge on a table and spreading the fabric on the plane while changing neither position nor attitude.

Note that fabric picked up from a table, and unfolded requires some way for placing the fabric flat.

3.2.2. Classifying planar unfolding elements

Planar unfolding in Section 3.2.1 is divided into unfolding using a single item following elements and using multiple items simultaneously.

(a) Inertia object

(b) Gravitational force

(c) Friction

(d) Contact with fluid (air, water, etc.)

(e) Handling

We studied (7) of 3.2.1, above, i.e., grasping an overlapped edge on a table and spreading the fabric on the plane while changing neither position nor attitude, i.e., a robot hand grasps an overlapping edge of fabric and unfolds it on a plane, a primitive motion consisting of handling under (e).

4. Method

4.1. Unfolding of Fabric by Handling

We assume that the robot grasps the fabric by nipping the edge and that the grasped position is one corner of the unfolded fabric (five if the fabric is a pentagon). Overlapping parts (hems) must be found but are usually difficult to detect using camera images alone in computer processing, even though humans do so easily. When the planar unfolded object, i.e., fabric, is input as a model beforehand --in the experiment, the model is generated by image input --and the robot searches for corner edges by predicting overlapping through partial pattern matching of overlapping fabric with the model. The corners are edges; likewise, predicted positions are edge corner points, the outline of overlapping fabric observed from above is a folded pattern, and the outline of unfolded fabric is an unfolded pattern. Conventional studies on partial pattern matching include searching for partially occluded objects. Few studies covered matching objects whose whole shape markedly changes. Because fabric is flexible and deforms, it cannot be matched against models highly accurately as with rigid objects. We adopted the following approach for unfolding: first, predicted points obtained through partial pattern matching are considered candidate points, and handling is conducted. If after a trial the point is considered erroneous, estimation is improved by repeating the process with other selected candidates, partially pattern-matching again, and improving prediction accuracy by adding information from other sensors.

4.2. Partial Pattern Matching and Edge Corner Prediction of Flexible Objects Using Outline Information

Outline information was represented by a 1-D array of data corresponding to the clockwise-arranged interior angles of the polygons representing the flexible object and segment length. [Figure 8]

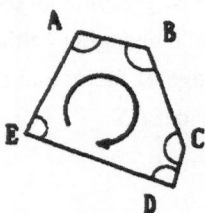

one dimensional data of outline information = { $\angle$ A,AB, $\angle$ B,BC, $\angle$ C,CD, $\angle$ D,DE, $\angle$ E,EA}

Figure 8: Examples of outline pattern and information for partial pattern matching

We assumed that the flexible object permits approximation by a polygon, and that at least one unfolded pattern (one edge corner or one side) is visible in the folded pattern. Overlapping corner edges are predicted through partial pattern matching as follows:

(1) Prepare unfolded pattern array WA from unfolded pattern array A (the number of elements is 2m assuming the object is an m-gon) and prepare folded pattern array WB from folded pattern array B (the number of elements is 2n assuming the object is an n-gon), where each array is 1-D (the number of elements is 2m+2n-2) More specifically, sequentially place elements of unfolded pattern array A into unfolded pattern array WA starting with the first element. Sequentially place each element of array A into array WA again starting with the (2m+l)th element, continuing this until all elements of array WA are filled with data. Prepare folded pattern array WB the same way.

WA(1) WA(2) .... WA(2m) WA(2m+1) .... WA(2m+2n-2)

↑ ↑ ↑ ↑ ↑

A(1) A(2) .... A(2m) A(1) .... A(2n-2)

(2) Compare each element of folded pattern array WB with each element of unfolded pattern array WA. A certain deformation is allowed in element matching because the compared object, i.e., fabric, is easily deformed. If the difference between corresponding folded and unfolded pattern elements is less than 10% of that

element of the unfolded pattern (i.e., if (WB(i))-WA(i)/WA(i)|<0.1) then a match is considered successful.

(3) Prepare a new array by placing the third and succeeding each element of un-folded pattern array A in unfolded pattern array WA in sequence, and match the new array against folded pattern array WA as in (2) above.

WA(1) WA(2) .... WA(2m) WA(2m+1) .... WA(2m+2n-2)

  ↑   ↑   ↑   ↑   ↑

A(3) A(4) .... A(2) A(3) .... A(2n)

(4) Prepare a new array by placing fifth and succeeding elements of unfolded pattern array A into the first and succeeding elements of unfolded pattern array WA in sequence, and match the new array against folded pattern array WB as in (2).

WA(1) WA(2) .... WA(2m) WA(2m+1) .... WA(2m+2n-2)

  ↑   ↑   ↑   ↑   ↑

A(5) A(6) .... A(4) A(5) .... A(2n+2)

Repeat until as many unfolded pattern arrays WA as the total number of corners in the folded pattern (n) are prepared, and match as in (2).

(5) Select the combination unfolded pattern array WA in which the largest num-ber of elements match consecutively with the candidate, i.e., that which is most similar to the folded pattern based on partial pattern matching. If the unfolded pat-tern completely coincides with the folded pattern, the number of elements that con-secutively match equals the number of array elements (2m+2n-2). The numbers of elements that consecutively match are the largest in multiple combinations. The sequence that will be obtained first under items (3) and (4) is designated the candi-date.

(6) Determine correspondence matching elements in the candidate pattern, i.e., determine which edge corners of the matching folded pattern correspond to edge corners of the unfolded pattern.

(7) Search for the unmatching edge corners in the unfolded pattern (point d in Figure 9), i.e., search for edge corners of the seam to be folded.

(8) Search for the segment closest to the edge corner selected under (7) (seg-ment BC in Figure 9), i.e., search for the part where the object seams to be folded.

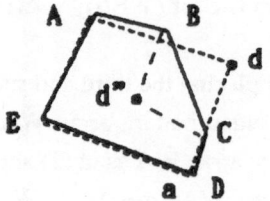

Figure 9: A point d of an estimated edge is the symmetrical point d with respect to the BC-line.

(9) Compute the position of the overlapping edge comer point (d" in Figure 9) assuming that the fabric is folded sharply

(10) The edge corner point farthest from the segment of folding is designated the first candidate edge corner point if multiple edge corner points are included in the overlapping part or multiple parts overlap.

Figure 9 shows the dotted lines that connect the candidate edge corner point with both end segments where fabric is folded.

4.3. Trial Pickup

Actual pickup is next. Following are conceivable causes of failure [Figure 10]:

(C1) The predicted point greatly deviates from the true position.

(C2) The fabric is folded so that the folded part is in the gap between the fabric and table.

(C3) A wrong point was designated as the candidate edge corner point due to partial pattern matching. (This can occur when a fold leads to an ambiguity in detecting the folded region due to the similarity in outline with another possibility.)

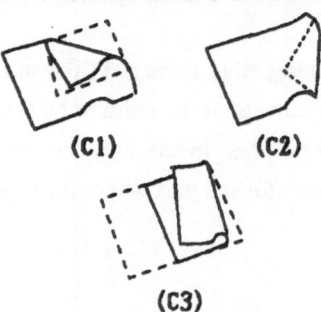

Figure 10: Examples of failed pickup. (C1) The estimated position of an edge corner has large error. (C2) A part is bent under itself. (C3) Incorrect partial pattern matching.

These failures must be coped with. Causes are determined by the following actions, and planar unfolding motion is tried again:

(Q0) Check if the robot has grasped an edge.

(Q1) Move the hand slightly toward where the fabric is unfolded, and try pickup.

(Q2) Determine changes in the height supposed edge part.

(Q3) Compute the edge corner point again using planar unfolding of another candidate obtained due to partial pattern matching, and try pickup.

Take measurements using tactile or visual sense or both in conducting motion under (Q0) through (Q2). The operation incorporating these motions in advance is trial pickup.

5. Experiment

5.1. Prototype of Flexible Object Handling

Our prototype system is shown in Figure.11. A CCD camera on the ceiling shows the overall view, including the robot. A mobile camera is also installed (there is a CCD camera at the end of the robot arm) so that the object and the robot motion can be observed more exactly. The arm has a flexible object handling effector (or hand). The experimental robot (Figure 4(d)) is equipped with a tactile sensor (Puma 260 Kawasaki Heavy Industries) and a camera installed on the ceiling (Toshiba CCD, IK-C 30). The hand is used as a tactile sensor and a fabric thickness check sensor by measuring finger tip distortion.

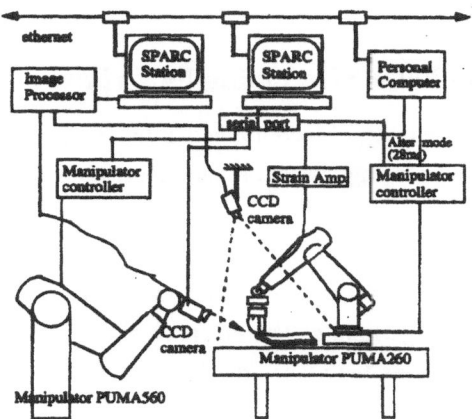

Figure 11: A prototype system for Handling Fabric

5.2. Fabric Unfolding Experiment

We studied how to predict the edge corner point of overlapping fabric using visual information and planar unfolding of fabric by basic cooperative sensing using the results of edge corner point prediction and tactile sensing as basic fabric handling. The fabric was plain, and its thickness known. The experiment was as follows, and (P2) through (P7) were repeated:

(P1) Obtain outline information by inputting an image of planar unfolded fabric.

(P2) The operator picks up the fabric once and places it within the camera's visual field differently from (P1) so fabric overlaps.

(P3) Input the overlapped fabric image and obtain outline information (supplement).

(P4) Predict the overlapped edge as in Section 3.2, using information obtained in (P1) and (P2).

(P5) Set the direction of finger motion to that in which the fabric is unfolded (d" to d in Figure.8), and move the hand to the predicted position above. The probability that the hand grasps the fabric is increased by grasping in the direction the fabric is to be unfolded in even if the predicted position of the edge corner point slightly deviates from the true position.

(P6) Lower the hand vertically, detect contact with the fabric on the table with the tactile sensor, and start pickup.

(P7) Determine whether the hand picked up the fabric based on the fabric thickness sensed by the tactile sensor If the hand has grasped the fabric properly, lift the edge vertically in the direction the fabric is to be unfolded in (toward d in Figure.8) on a circular trajectory of radius 25 mm, transfer it horizontally, and release it 30 mm before the target point.

5.3. Results of the Fabric Unfolding Experiment

This experiment used the object-oriented robot programming language Eus-Lisp[19], developed considering system integration for all types of computer processing except low-level image processing. Results on partial pattern matching and prediction of overlapping corner edge are shown in Figure 12. We verified that partial pattern matching is effective for both convex and concave polygons [Figure 12]. The visual input image field covers an area about 47 cm wide on this side, 46 cm deep, and 51 cm wide on the other side; horizontal image resolution is 512 pixels and vertical 480 pixels. Error for one pixel is equivalent to about 1 mm.

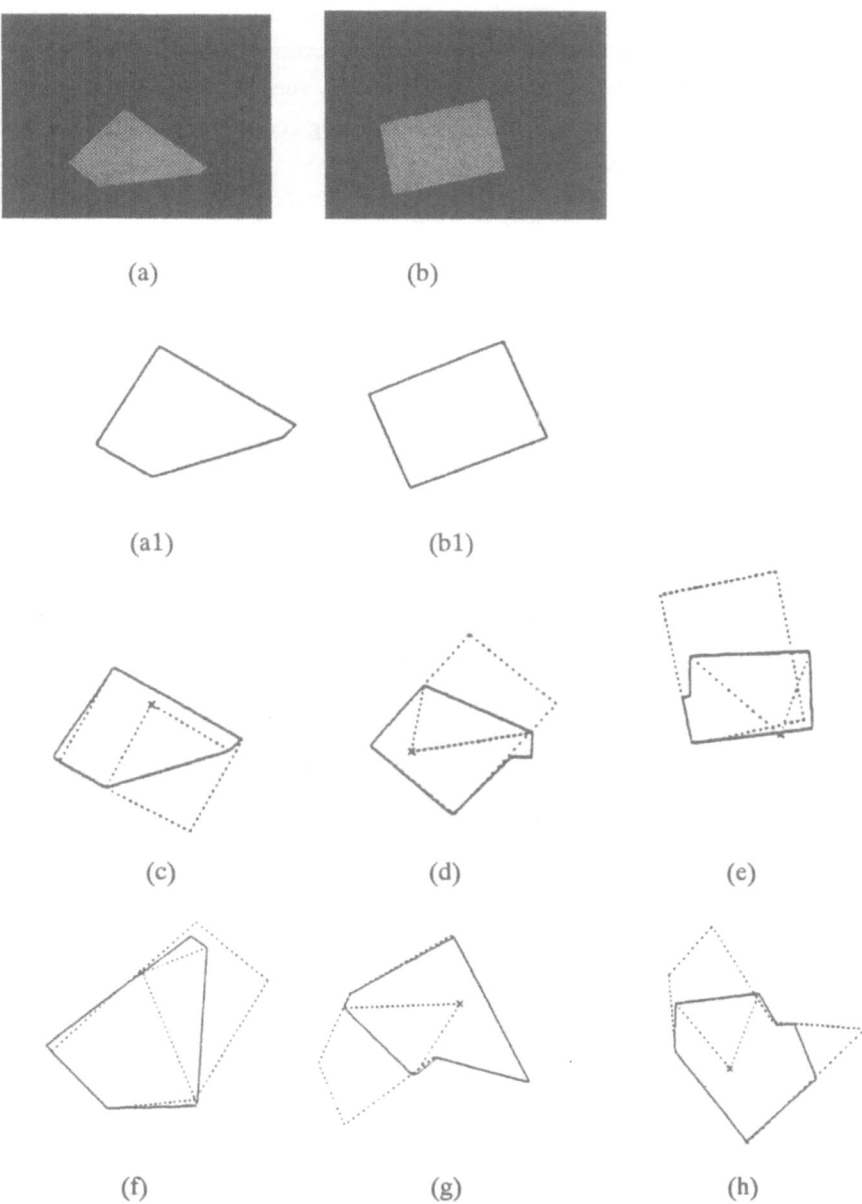

Figure 12: (a) Raw image of folded part, (b) raw image of unfolded part, (a1,b1) top view of polygonal approximations after image processing (a) and (b), (c) partial pattern matching between (a1) and (b1), (d-h) other samples of partial pattern matching, (e) partial pattern matching in error (C3) in Figure 10, (h) partial pattern matching in case that 2 parts are folded.

Edge corners are not necessarily predicted accurately, because visual information is affected by error accuracy, and the distortion accumulates every time the fabric is placed on the table. The experiment was basically successful in unfolding the fabric except when the robot was unable to grasp the edge corner due to the reasons given in C1 through C3 in Section 4.3. [Figure 13]

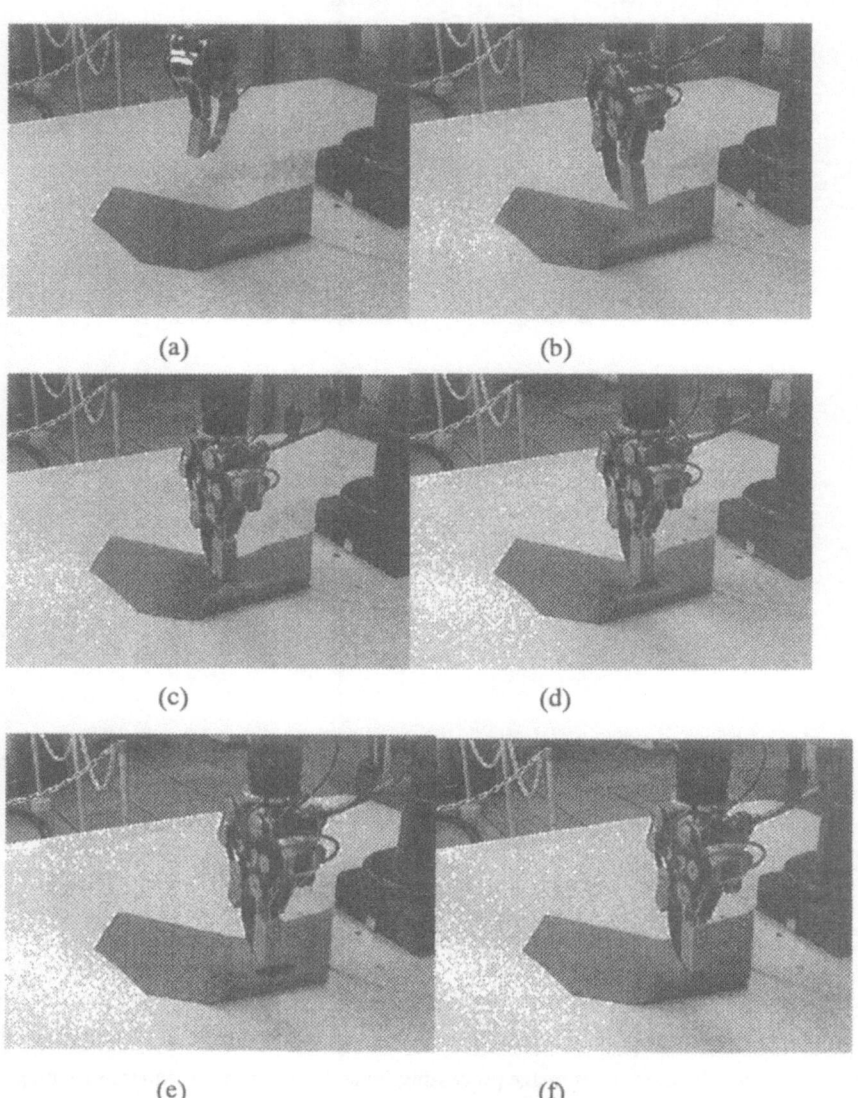

(a)

(b)

(c)

(d)

(e)

(f)

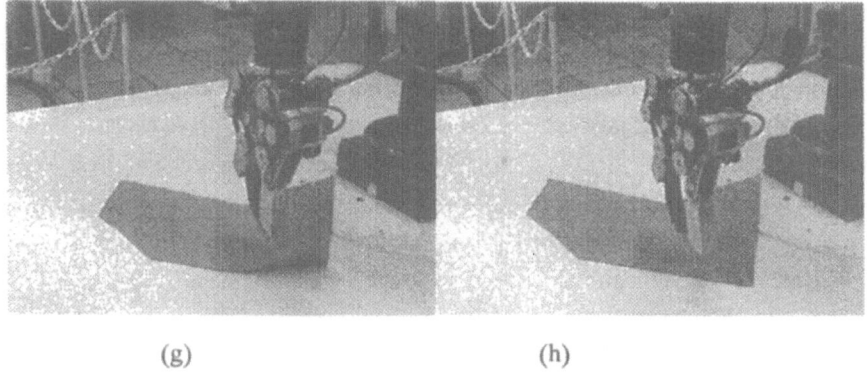

(g)   (h)

Figure 13: Sequence of picking up and unfolding. (a)(b) motion to position over candidate edge,(c) decision on whether a piece is held correctly or not after detection of height of the piece, (d-f) transferring the corner edge, (g-h) the part is released.

An edge corner point is usually difficult to find using visual information alone, as is finding appropriate grasp and unfolding motions without changing the position of the fabric using tactile sense information alone, because the shape tends to change. Fabric unfolding was realized by combining visual sense information for predicting an edge corner and tactile sense information for determining whether the robot hand came into contact with and grasped the fabric. Allowance for deformation is large in matching corresponding parts in unfolded and folded patterns if the fabric is wrinkled and creased. More complex cooperative visual and tactile sensing will be required.

5.4. Experimental Procedure and Result Measurement of Changes in Fabric's Edge Height

In Section 4.3, if the robot fails to pick up fabric, we studied solving the problem in (C2) by experimenting with the overlapping fabric, discriminated by using information on the height of fabric obtained by touching it [(Q2) in Section 4.3]. The contact force when the robot hand touches the object is sensed by measuring the robot finger tip deformation in contact using a phosphor bronze plate sensor. Changes in the fabric surface height were measured by making the robot hand lightly and successively stroke the surface of the overlapped fabric on the table at the same strength while it is moved, and recording the movement transmitted to the robot hand during motion. The vertical contact force applied to the fabric was about 0.02 N, found by loading fabric on an electronic force balance (sartorius analytical

balance: model L420S able to measure force as small as 0.001 g) and stroking the
fabric with the robot hand the same way.

A movement instruction was given to the robot every eight control cycles so it
moves unfailingly to the instructed position. A single control cycle of the robot is
about 28 ms. The trajectory was computed based on the movement instruction
given to the robot. In this experiment, two types of fabric -- a piece of rayon fabric
0.2 mm thick and a piece of wool 0.75 mm thick -- were folded [Figure 14], and
the robot hand stroked them in directions P2 and P3 [Figure 15].

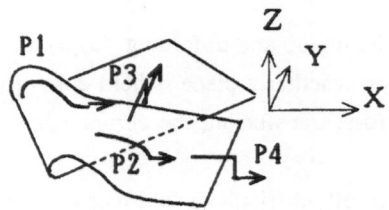

Figure 14: Detection difference of fabric surface height. P1 is higher. Although there are
steps in P2,P3,andP4, it is possible to turn over the edge in P3 or P4, but not in P2.

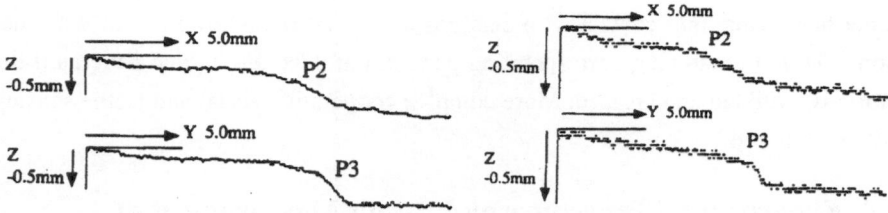

Figure 15:The plot robot hand when a fabric surface was stroked and the surface was sensed.

The thicker the fabric, the greater the difference between exhibited height when
the fabric is stroked toward P3. When the measurement is toward P3, the downward
step to the surface of the bottom fabric occurs when the robot hand strokes the top
of fabric. When measurement is toward P2, the slope step is gentle, because the top
fabric covers the bottom fabric. Thus, measurement of changes in folded fabric
height taken by stroking the surface with the robot hand enables us to determine
whether the fabric is folded upward or downward, if the fabric is 0.2mm or more
thick.

Because the finger tip width of the robot used in the experiment is 2 cm, the difference in the height of fabric is difficult to determine clearly unless the finger tip strokes the fabric perpendicular to the edge of fabric. Irregular fabric is measured using tactile sense information, and the prediction that fabric overlaps is verified while the finger tip strokes the fabric surface perpendicular to the edge predicted based on visual sense information. If edge prediction is correct, we can also determine whether fabric is folded upward or downward. As this experiment indicates, cooperative visual and tactile sensing is basically active because it includes and tactile sensing motion ; this sensing is vital to handling, which includes many uncertainties.

6. Conclusion

I introduced several handling devices developed under the automated sewing system project. Those new techniques were now used in products & machines. However handling fabric is still difficult problem. I discussed important issues and the need for cooperative sensing to handle flexible objects. This involves the first prototype for unfolding fabric, and yielded the following results :

1 : Partial pattern matching was done by using a polygon to approximate the outline of a piece of fabric obtained by visual sense information, then using the outline information of that polygon.

2 : The robot was able to grasp, confirm that it grasped fabric using the tactile hand, and unfold fabric to pick up the corner edge based on partial pattern matching, assuming that the comer edge point is overlapping at the same position as when the fabric was folded in a similar way to origami (paper folding). However, the robot failed in grasping the corner edge properly when the predicted position had a large error or was wrong. Failure was caused mainly by three types of error.

3 : For errors in determining whether fabric is folded upward or downward, one of three types of prediction errors, the experiment verified that the robot is able to distinguish between overlapping states by stroking the fabric 0.2 mm or thicker.

Possible future studies in cooperative sensing are :

*Error recovery at pickup

*Partial pattern matching that includes curvilinear parts

*Extraction of 3-D information by visual sense information

*Design of robot hands and 3-D handling

Previous studies on fabric handling were about making robots conduct specific work in a limited environment. This study targets on technology for versatile handling of fabric, and is accompanied by concurrent experiments actually handling fabric. Handling flexible and soft materials is one of necessary techniques not only for industrial production system and but also for service and assisted robots. Therefore, theses research can be more practical as research projects for coexistence of man and robots included humanoid robots advance

References :

1. Kawachi 1979 Development of the Material Handling Technology for Parts Sewing (1) *J Textile & Fiber Engineering* Vol 32 No 6 pp310-318

2. Kawachi 1979 Development of the Material Handling Technology for Parts Sewing (2) *J Textile & Fiber Engineering* Vol 32 No 6 pp310-318

3. J. K. Parker, R. Dubey, F. W. Paul and R. J. Becker, 1982 Robotic Fabric Handling for Automated   Garment Manufacturing *Trans. of the ASME J Engineering for Industry* pp1-6

4. D. R. Kemp, G. E. Taylor, P. M. Taylor and A. Pugh 1983 A Sensory Gripper for Handling Textiles *13th Int. Sump. on Industrial Robots and Robots* Vol 2 pp18-23-18-33

5. E. Torgerson and F. W. Paul 1987 Vision Guided Robotic Fabric Manipulation for Apparel Manufacturing 1987 *IEEE Int. Conf. on Robotics and Automation* Vol 2 pp1196-1202

6. D. Gershon and I. Porat 1988 Vision Servo Control of a Robotic Sewing System 1988 *IEEE Int. Conf. on Robotics and Automation* Vol 3 pp1830-1835

7. P. M. Taylor and S. G. Koudis 1988 The Robotic Assembly of Garments with Concealed Seams 1988 *IEEE Int. Conf. on Robotics and Automation* Vol 3 pp1836-1838

8. P. M. Taylor 1990 Sensory Robotics for the Handling of Limp Materials *Springer-Verlag* pp141 - 192

9. G. Schulz 1991 Grippers for Flexible Textiles 1990 *IEEE Int. Conf. on Advanced Robotics* Vol 1 pp759-764

10. Shigeki, Iwane, Nakata, Takahashi, Nakazawa and Kudo 1992 A Study on the Positioning and Grasping of Fabric by Robot Hand 1992 *JSME* No920-78 pp497-499

11. P.M. Taylor, D.M. Pollet and M.T. GrieBer 1994 Analysis and Design of Pinching Grippes for the Secure Handling of Fabric Panels *EURISCON '94* Vol 4 pp1847-1856

12. K. Parashidis, N. Fahantidis, V. Vassiliadis and et al 1995 A Robotic System for Handling Textile Materials *1995 IEEE Int. Conf. on Robotics and Automation* Vol 2 pp1769-1774

13. E. Ono, H. Ichijo, N. Aisaka and H. Akami 1990 Robot Hand with Sensor for Handling Cloth *1990 IEEE Int. Workshop on Intelligent Robots and Systems* Vol 2 pp995-1000

14. E. Ono, H. Ichijo and N. Aisaka 1991 Robot Hand for Handling Cloth *1991 IEEE Int. Conf. on Advanced Robotics* Vol 1 pp769-774

15. E. Ono, H. Ichijo and N. Aisaka 1992 Flexible Robotic Hand for Handling Fabric Pieces in Garment Manufacture *Int. J Clothing Sci. and Technology* Vol 4 No 5 pp16-23

16. E. Ono, H. Okabe, H. Akami and N. Aisaka 1989 Robot Hands with Sensors for Handling Fabric *J the Textile Machinery Society of Japan* Vol 42 No 6 pp41-52

17. E. Ono, S. Nishikawa, H. Ichijo and N. Aisaka 1992 New Robot Hand for Cloth Handling *J Sen-i Gakkaishi* Vol 48 No 9 pp501-506

18. E. Ono, N. Kita and S. Sakane 1995 Strategy for Unfolding a Fabric Piece by Cooperative Sensing of Touch and Vision *1995 IEEE/JSR Int. Conf. on Intelligent Robots and Systems* Vol 3 pp441-445

19. T. Matsui and M. Inaba 1990 EusLisp: An Object-based Implementation of Lisp *J Information Processing* Vol 13 No 3 pp.327-328

Appendix:

· Outline information was obtained as follows:

The following preparation was made for 256 gradations of 512x512 pixels of input image using general-purpose software for image processing, Khoros:

1: Data were binarized after reducing the effects of objects other than the fabric on image processing by taking the differences between the images before and after placing the fabric on the table. The threshold value used for binarizing was determined by humans based on fabric colors and the table.

2: The absolute respective difference between binarized image (A) and images (B) and (C) was obtained by shifting image (A) toward X and Y by one pixel. The logical sums of two images, i.e., (|A-B| or |A-C|), were set as the outline object and approximated by straight lines.

Then, the following computation was done using Euslisp:

3: Comparison of the folded pattern and the unfolded pattern requires the knowledge of the object shape viewed from above. To satisfy this requirement, the X and Y coordinates of the end points of the approximate straight lines used were transformed into 3-D space based on a prior calibration of the fixed camera. Note that the table surface was horizontal in the robot's coordinate frame.

4: The straight line outline obtained above (a single side consisted of multiple straight lines in many cases) was approximated by a polygon. 1-D arrays were prepared by arranging in sequence the finally obtained straight lines and the interior angles between these straight lines.

Chapter 4
Collaborative Systems

Section 4.1

Manipulation of Sheet Metal by Dual Manipulators

H. Yoshida and K. Kosuge

Abstract. In this paper, we propose a control algorithm of dual manipulators handling flexible metal sheet: we discuss how to bend metal sheet and how to manipulate it. First, we derive the relationship between the static deformation of the sheet metal and bending moments exerted on the sheet using the Lagrange equation based on its finite-element model. We then design a control algorithm by which the motion of the manipulated sheet is controlled using the resultant force applied to the sheet. Since the rigidity of the deformed sheet is not uniform, that is, the stiffness of the deformed sheet may depend on the direction along which external force is applied, we cannot use the compliance of the flexible sheet for tasks involving interactions between the sheet and its environment. The proposed control algorithm designed so that the apparent impedance of the manipulated sheet is specified. The experimental results using industrial robots illustrate the validity of the proposed control system.

1. Introduction

Manipulation of flexible materials is one of the difficult processes on the production line in the manufacturing system. Even now many processes are still done by the skillful workers. So handling of flexible materials attracts a great deal of attention.[2]-[5] As flexible materials are controlled based on the strict model, it is difficult to model complicated flexible materials. So we proposed a control of a metal sheet based on the finite element techniques that is applicable to modeling complicated flexible materials. When we bend a metal sheet, we can get different rigidity on each direction shown in figure 1. After getting the proper rigidity on each direction, we want to press the deformed metal sheet along the rigid direction shown in figure 1(a). Considering the external force from the environment, we proposed a rigid object around the representative point of it.[1] So we present a method to control each manipulator around the desired representative point of a metal sheet.

This paper is organized as follows: in the next section, we consider the problem of handling a metal sheet; the third section introduces the modeling of a metal sheet based on the finite element techniques; in the fourth section, we propose a control algorithm of a metal sheet based on the two impedance controlled manipulators; experimental testing by dual manipulators are presented in the fifth section; finally, the results of this paper are summarized.

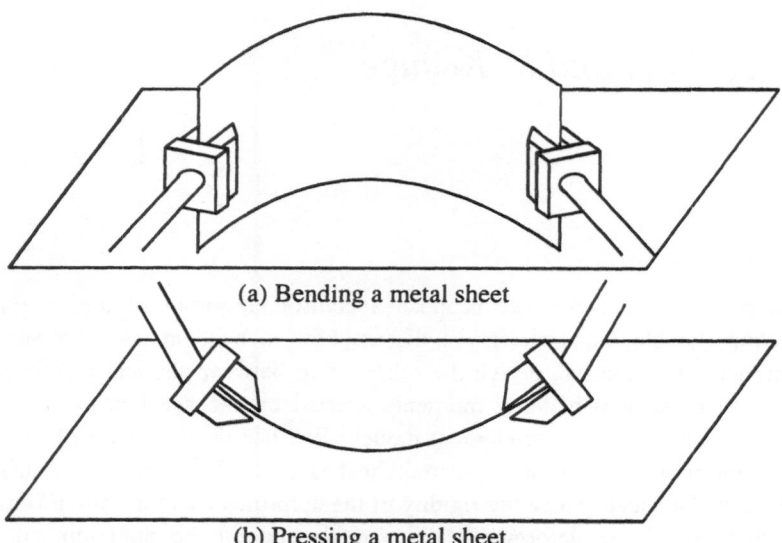

(a) Bending a metal sheet

(b) Pressing a metal sheet

Figure 1: Examples of bending and pressing a metal sheet

2. Deformation of the flexible materials

We consider a metal sheet as a examples of the flexible materials. To bend and press a deformed metal sheet shown in figure 1(a), we consider two problems as follows;
(1) the relation between moments to deform a metal sheet and the displacement of it
(2) to control each manipulator to perform the desired motion to external forces from environment

The problem pointed out in (1) includes the following deformations: i) expansion and contraction along the axial direction, ii) bending, and iii) torsion. In this research, we analyze the bending of a metal sheet based on the finite element model that is made good use of engineering. We cope with the problem (2) by controlling each manipulator around a representative point of a flexible metal sheet.

Considering the deformation of a flexible metal sheet, we control it under following assumption;
1) each manipulator grasps a metal sheet firmly.
2) the bending of a metal sheet is considered within the elastic deformation.
3) a metal sheet is bent statically, after it was grasped.

3. Modeling of a flexible metal sheet

In order to analyze a metal sheet based on the finite element model, we attach a coordinate frame to the grasped point of the left arm, with X axis aligned with the metal sheet when it is not bent shown in figure 2. Here we use the method presented in [6]. We divide a metal sheet into the appropriate N sections with the same interval h. They are defined 1, 2,..., N from the left side. Nodes are defined to investigate the motion of each element and they are defined 0, 1, 2, ..., N from left side. The displacement of the each node is $u_0(t), u_1(t), ..., u_{N-1}(t), u_N(t)$, where $u_1(t), ..., u_{N-1}(t)$ are unknown. As a metal sheet is grasped firmly by dual manipulators, $u_0(t), u_N(t)$ are known.

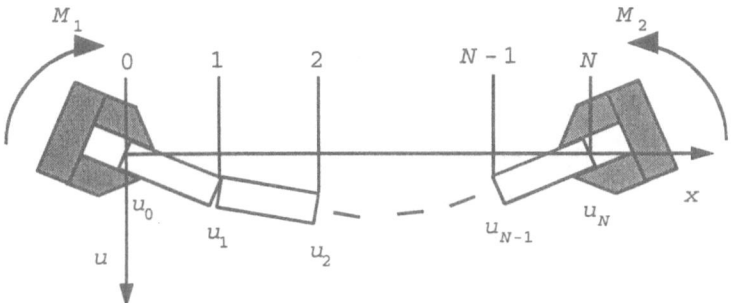

Figure 2: Analysis using finite element techniques

Next we consider an element. We set i-1 and i at the edges of an each element. It is necessary to take u and θ as degrees of freedom at each node of an element. Therefore, the element with two nodes has a total of four degrees of freedom shown in figure 3. The displacement function $u_{i-1,i}$ can thus be represented by a polynomial with four constraints, namely

$$u_{i-1,i}(x_i,t) = \alpha_0 + \alpha_1 x_i + \alpha_2 x_i^2 + \alpha_3 x_i^3 \qquad (1)$$

from the boundary condition

$$u_{i-1,i}(0,t) = u_{i-1}(t)$$

$$\partial u_{i-1,i}(x_i,t)/\partial x_i \,|_{x_i=0} = \theta_{i-1}(t) \qquad (2)$$

$$u_{i-1,i}(h,t) = u_i(t)$$

$$\partial u(x_i,t)_{i-1,i}/\partial x_i \,|_{x_i=h} = \theta_i(t) \qquad (3)$$

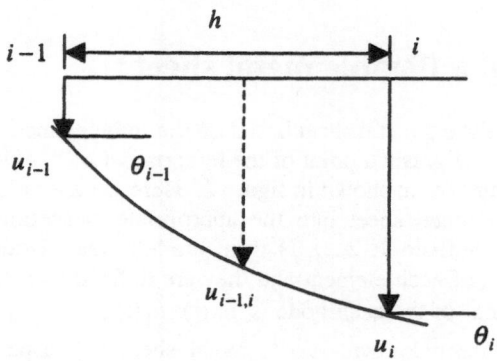

Figure 3: Approximation by a cubic function

Substituting (2), (3) into (1), we can get following equation

$$u_{i-1,i}(x,t) = \left[1 - 3(x_i/h)^2 + 2(x_i/h)^3\right]u_{i-1}(t) +$$
$$\left[(x_i/h) - 2(x_i/h)^2 + (x_i/h)^3\right]h\theta_{i-1}(t) + \qquad (4)$$
$$\left[3(x_i/h)^2 - 2(x_i/h)^2\right]u_i(t) +$$
$$\left[-(x_i/h)^2 + (x_i/h)^3\right]h\theta_i(t)$$

This equation can be expressed in the form

$$u_{i-1,i}(x_i,t) = N(x_i)_i^T v(t)_i \qquad (5)$$

where

$$N(x_i)_i = \begin{bmatrix} n_1(x_i) \\ n_2(x_i) \\ n_3(x_i) \\ n_4(x_i) \end{bmatrix}, \qquad v(t) = \begin{bmatrix} u_{i-1}(t) \\ h\theta_{i-1}(t) \\ u_i(t) \\ h\theta_i(t) \end{bmatrix} \qquad (6)$$

The displacement function $N(x_i)_i$ in (6) are given by

$$n_1(x_i) = 1 - 3(x_i/h)^2 + 2(x_i/h)^3$$
$$n_2(x_i) = (x_i/h) - 2(x_i/h)^2 + (x_i/h)^3$$
$$n_3(x_i) = 3(x_i/h)^2 - 2(x_i/h)^2 \qquad (7)$$
$$n_4(x_i) = -(x_i/h)^2 + (x_i/h)^3$$

The kinetic energy of a small increment dx_i, is $(1/2)\rho A(\partial u/\partial x)^2 dx_i$. The kinetic of the complete element is therefore

$$K_i = \frac{1}{2}\int_0^h \rho A\left(\frac{\partial u}{\partial t}\right) dx_i \qquad (8)$$

where A is an area of the cross-section, and ρ is the mass per unit volume of the element.

And the strain energy stored in the element is given by

$$P_i = \frac{1}{2}\int_0^h EI_z\left(\frac{\partial^2 u}{\partial x_i^2}\right) dx_i \qquad (9)$$

where I_z is the second moment of area of the cross-section about the z axis and E is Young's modulus for the material.

We also consider the virtual displacement ∂W_i in the element to calculate the virtual work. The virtual work done by the external force $f(x_i,t)$ is

$$\partial W_i = \int_0^h f(x_i,t)\partial u_{i-1,i}(x_i)dx_i \qquad (10)$$

Substituting the displacement expression (5) into the kinetic energy gives

$$K_i = \frac{1}{2}\int_0^h \rho A v(t)_i^T N(x_i)_i N(x_i)_i^T v(t)_i dx_i$$

$$= \frac{1}{2}v(t)_i^T m_i v(t)_i \qquad (11)$$

Therefore the element inertia matrix is given by

$$m_i = \int_0^h \rho A N(x_i)_i N(x_i)_i^T dx_i \qquad (12)$$

Substituting for the function (12) from (7) and integrating gives

$$m_i = \frac{\rho Ah}{420}\begin{bmatrix} 156 & 22 & 54 & -13 \\ 22 & 4 & 13 & -3 \\ 54 & 13 & 156 & -22 \\ -13 & -3 & -22 & 4 \end{bmatrix} \qquad (13)$$

Substituting the displacement expression (5) into the strain energy (9) gives

$$P_i = \frac{1}{2}\int_0^h EI_z \{v(t)\}_i^T \left\{\frac{d^2 N(x_i)}{dx_i^2}\right\}\left\{\frac{d^2 N(x_i)}{dx_i^2}\right\}_i^T \{v(t)\}_i dx_i$$

$$= \frac{1}{2}\{v(t)\}_i^T k_i \{v(t)\}_i \qquad (14)$$

The element stiffness matrix is therefore

$$k_i = \int_0^h EI_z \left\{\frac{d^2 N(x_i)}{dx_i^2}\right\}\left\{\frac{d^2 N(x_i)}{dx_i^2}\right\}_i^T dx_i \qquad (15)$$

Substituting for the function (15) from (7) and integrating gives

$$k_i = \frac{EI}{h^3} \begin{bmatrix} 12 & 6 & -12 & 6 \\ 6 & 4 & -6 & 2 \\ -12 & -6 & 12 & -6 \\ 6 & 2 & -6 & -4 \end{bmatrix} \tag{16}$$

The virtual work done by the external force becomes, after substituting (5) into (11),

$$\partial W_i = \int_0^h f(x_i,t)N(x_i)_i \partial v_i dx_i \tag{17}$$

where

$$\int_0^h f(x_i,t)N(x_i)_i dx_i = f_i, \quad f_i = \begin{bmatrix} f_{i-1} \\ m_{i-1} \\ f_i \\ m_i \end{bmatrix} \tag{18}$$

Lagrange's equations now take the form

$$\frac{d}{dt}\left(\frac{\partial L}{\partial \dot{v}_i}\right) - \left(\frac{\partial L}{\partial v_i}\right) = f_i \tag{19}$$

where

$$L = K_i - P_i \tag{20}$$

Substituting equations (12),(15) and (17) into (19) gives

$$m_i v_i + k_i v_i = f_i \tag{21}$$

We only consider the static motion of the flexible metal sheet, so we can neglect the kinetic energy. The equation (21) transforms the following equation

$$k_i v_i = f_i \tag{22}$$

So the static motion of the metal sheet is expressed as follows

$$KV = F \tag{23}$$

where

$$K = \frac{EI}{h^3} \begin{bmatrix} 12 & 6 & -12 & 6 & \cdots & \cdots & \cdots & \cdots & \cdots & 0 \\ 6 & 4 & -6 & 2 & & & & & & \vdots \\ -12 & -6 & 24 & 0 & & & & & & \vdots \\ 6 & 2 & 0 & 8 & & & & & & \vdots \\ \vdots & & & & \ddots & & & & & \vdots \\ \vdots & & & & & \ddots & & & & \vdots \\ \vdots & & & & & & 24 & 0 & -12 & 6 \\ \vdots & & & & & & 0 & 8 & -6 & 2 \\ \vdots & & & & & & -12 & -6 & 12 & 6 \\ 0 & \cdots & \cdots & \cdots & \cdots & \cdots & 6 & 2 & -6 & 4 \end{bmatrix} \tag{24}$$

$$V = \begin{bmatrix} u_0 & h\theta_0 & u_1 & h\theta_1 & \cdots & u_{N-1} & h\theta_{N-1} & u_N & h\theta_N \end{bmatrix}$$

$$F = \begin{bmatrix} f_1 & \frac{m_1}{h} & f_2 & \frac{m_2}{h} & \cdots & f_{N-1} & \frac{m_{N-1}}{h} & f_N & \frac{m_N}{h} \end{bmatrix}$$

For example, a metal sheet is divided into 6 elements. Each manipulator supports the flexible metal sheet rigidly in a horizontal line.

$$u_0 = u_N = 0 \tag{25}$$

Let us calculate the displacement and radian of each element when each manipulator adds the moment M, $-M$ shown in figure 2.

$$u_{max} = \frac{N^2 M h^2}{8EI} \tag{26}$$

$$\theta_0 = \frac{NMh}{2EI}, \quad \theta_N = -\frac{NMh}{2EI} \tag{27}$$

So we can guess the displacement of each element of a metal sheet.

4. Design of the control algorithm

4.1 Strategies for manipulating a metal sheet
        We deal with a problem of bending a metal sheet by dual manipulators shown in figure 1(a). After bending a metal sheet, we control the dual manipulators around the representative point of a metal sheet shown in figure 4. By realizing an apparent compliance of a metal sheet, we control each manipulator to perform the desired motion to external forces from the environment. Deformation of a metal sheet is controlled by internal moments which are not related to the motion of a metal sheet. So the control algorithm is applicable to bending and pressing a metal sheet.

4.2 Bending control of a metal sheet
        Let us consider a problem of bending a metal sheet supported by dual manipulators, when internal moments M, $-M$ are exerted on the edges of the metal sheet shown in figure 5. When the Arm1 adds the moment M at the end-effector, we assume that the desired radian of the deformed metal sheet at the end effector is $4u_{max} / Nh$. From the expression (27), the moment to deform a metal sheet at the end-effector of Arm1 is given by

$$M_d = \frac{2EI}{Nh} \theta_0^d \tag{28}$$

The control input to deform a flexible metal sheet is given by

$$M = \frac{2EI}{Nh} \theta_0^d + k_I \int_0^t \left(\theta_0^d - \theta_0 \right) dt \quad (k_I > 0) \tag{29}$$

where EI is a constant of integration
        Since the motion of the end-effectors is symmetric, the moment exerted on Arm2 at the end-effector is $-M$. So we control the Arm2 to satisfy following equation

$$M_d = -\frac{2EI}{Nh} \theta_0^d \tag{30}$$

Substituting equations (28) into (29) gives and differentiate

$$\dot{e} + \frac{k_I N h}{2EI} e = 0 \tag{31}$$

where

$$e = \theta_0^d - \theta_0 \tag{32}$$

4.3 Apparent mechanical impedance of a metal sheet to the external force

Let the representative point of a metal sheet be maximum displacement of a metal sheet. So we estimate the displacement of the representative point from the equation (26). After estimating the maximum displacement of a metal sheet, we control each manipulator around the maximum displacement of it shown in figure 4.

We assume that a metal sheet does not deform along the rigid direction. By controlling each manipulator around the representative point of it, we press a deformed metal sheet.

Let the mechanical impedance of a metal sheet to the external force be expressed by the following equation

$$M_0 \Delta \ddot{x} + D_0 \Delta \dot{x} + K_0 \Delta x = F^{ext} \tag{33}$$

where Δx is the deviation of the manipulated metal sheet from the desired trajectory of the representative point of it, F^{ext} is the external force applied to a metal sheet around the representative point of it, and M_0, D_0, K_0 is positive definite matrix.

The external force f_i^{ext} is expressed the following equation

$$f_i^{ext} = f_i - (f_i^d + f_i^0) \tag{34}$$

where f_i is the sensor force of i-th manipulator, f_i^d is the desired force to bend a metal sheet, f_i^0 is the load of a sheet.

Let the mechanical impedance of the i-th manipulator around the desired compliance center of a metal sheet be expressed by

$$M_i \Delta \ddot{x}_i + D_i \Delta \dot{x}_i + K_i \Delta x_i = f^{ext} \tag{35}$$

where f^{ext} is the external force applied to the i-th manipulator around the representative point of a metal sheet, and Δx_i is the deviation of the representative point of the i-th manipulator, and M_i, D_i, K_i is positive definite matrix.

We assume that each arm grasps a metal sheet firmly and no relative motion exists between the metal sheet and each arm.

$$x = x_i \tag{36}$$

Let us specify the external force shared by each arm as follows:

$$f_i^{ext} = \rho_i F^{ext} \tag{37}$$

Concerned with the external force applied to the metal sheet, the following relation holds:

$$\sum_{i=1}^{2} \rho_i = 1 \quad (\rho_i > 0) \tag{38}$$

and

$$\sum_{i=1}^{2} f_i^{ext} = F^{ext} \tag{39}$$

From equations (33)•(39), we obtain the mechanical impedance of each arm, which realizes the desired mechanical impedance of the manipulated sheet expressed by equation (35), as follows:

$$M_i = \rho_i M_0$$
$$D_i = \rho_i D_0 \tag{40}$$
$$K_i = \rho_i K_0$$

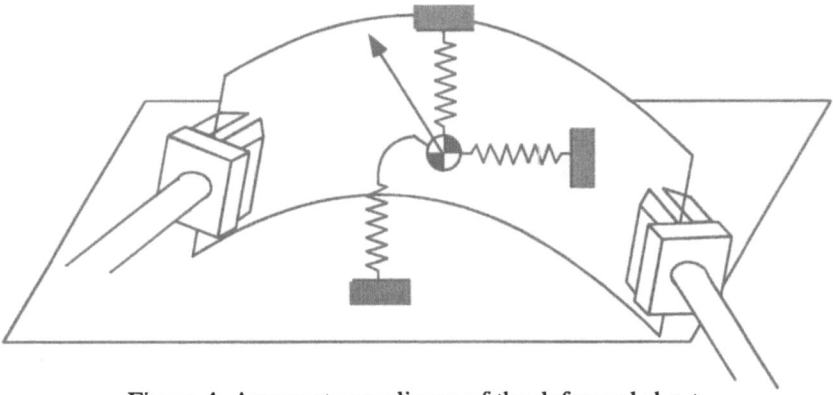

Figure 4: Apparent compliance of the deformed sheet

5. Experimental Study

5.1 Experiment of bending a metal sheet

We applied the proposed control algorithm of bending for the dual industrial manipulators system (Nachi, 7603-APJ), each of which has six degrees of freedom. The force sensor is mounted at the wrist of each manipulator for measuring the forces and moments exerted on the end-effector. The end-effectors are used to bend an aluminium sheet shown in figure 11. VxWorks was used to implement the control algorithm. Sampling rate was 500[Hz]. From the equation (26), the relation between the moment and the maximum displacement is $15 \times 10^{-3}[m]$. So the appropriate moment to realize is

$$M_d = \frac{8EIu_{max}}{N^2 h^2} \tag{41}$$

From the equation (27), the angles applied at a grasped point is

$$\theta_0 = \frac{NMh}{2EI}, \quad \theta_N = -\frac{NMh}{2EI} \tag{42}$$

Substituting (41) into (42) gives

$$M_d = \frac{2EIu_{max}}{9h^2} \tag{43}$$

The following results are obtained by controlling each manipulator to satisfy equation (29). In the experiment, the desired Euler angles of the two end-effectors are $13°, -13°$. Figure 8, 9 show the Euler angles of the two manipulators. Table1 shows the impedance parameters of each manipulator.

5.2 Experiment of pushing the deformed metal sheet along the direction of high stiffness

In the following experiment, we press the deformed metal sheet along the direction of high rigidity shown in figure 6,7. Each manipulator is controlled around the representative point of a metal sheet under the condition of equation (32). At first, each manipulator grasps a metal sheet firmly to keep Euler angles $80°$ and $-100°$. Figure 10, 11 show the Euler angle of each manipulator when a metal sheet is pressed. The desired Euler angles of the two end-effectors are $90°$ and $-90°$. Each end-effector rotates the representative point of the metal sheet.

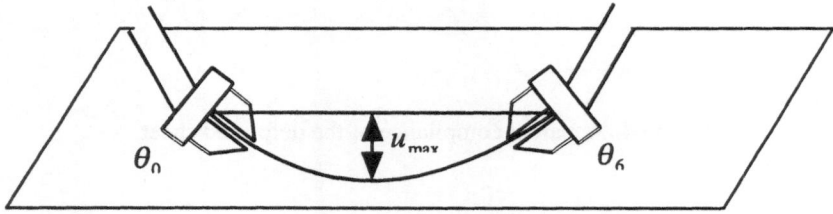

Figure 5: Bending a metal sheet

	$M_i[kg]$						$D_i[Ns/mm]$						$K_i[N/mm]$					
Arm 1	10	0	0	0	0	0	40	0	0	0	0	0	30	0	0	0	0	0
	0	10	0	0	0	0	0	40	0	0	0	0	0	30	0	0	0	0
	0	0	10	0	0	0	0	0	40	0	0	0	0	0	30	0	0	0
	0	0	0	1	0	0	0	0	0	2.5	0	0	0	0	0	1	0	0
	0	0	0	0	1	0	0	0	0	0	2.5	0	0	0	0	0	1	0
	0	0	0	0	0	1	0	0	0	0	0	2.5	0	0	0	0	0	1
Arm 2	10	0	0	0	0	0	40	0	0	0	0	0	30	0	0	0	0	0
	0	10	0	0	0	0	0	40	0	0	0	0	0	30	0	0	0	0
	0	0	10	0	0	0	0	0	40	0	0	0	0	0	30	0	0	0
	0	0	0	1	0	0	0	0	0	2.5	0	0	0	0	0	1	0	0
	0	0	0	0	1	0	0	0	0	0	2.5	0	0	0	0	0	1	0
	0	0	0	0	0	1	0	0	0	0	0	2.5	0	0	0	0	0	1

Table 1: Impedance parameters of each manipulator

171

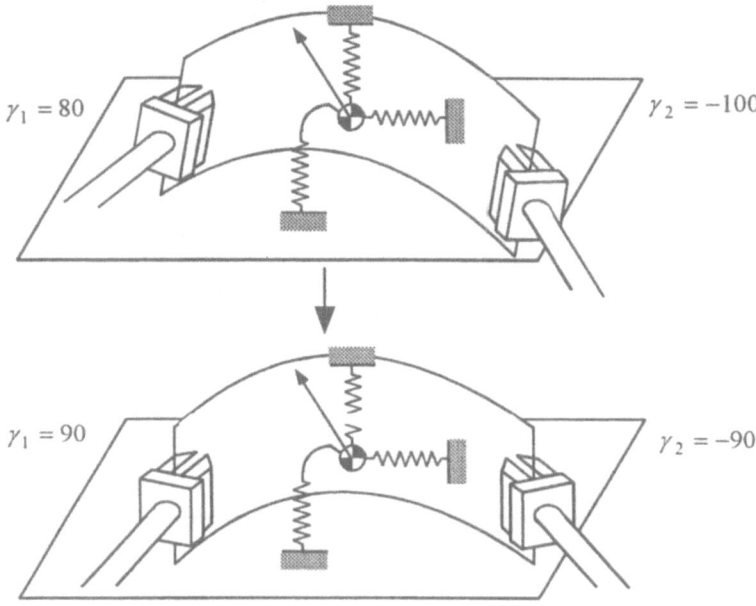

$\gamma_1 = 80$ $\gamma_2 = -100$

$\gamma_1 = 90$ $\gamma_2 = -90$

Figure 6: Pressing a deformed metal sheet

Figure 7: Experimental system

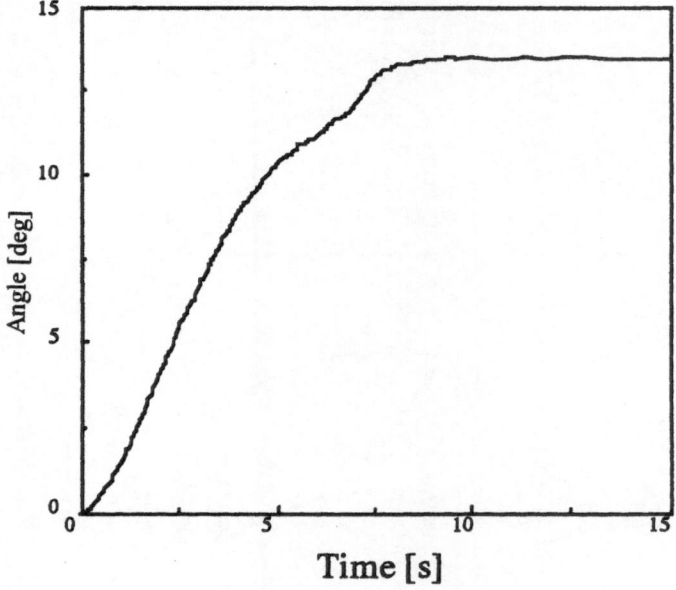

Figure 8: Euler angle θ_0 (Arm1)

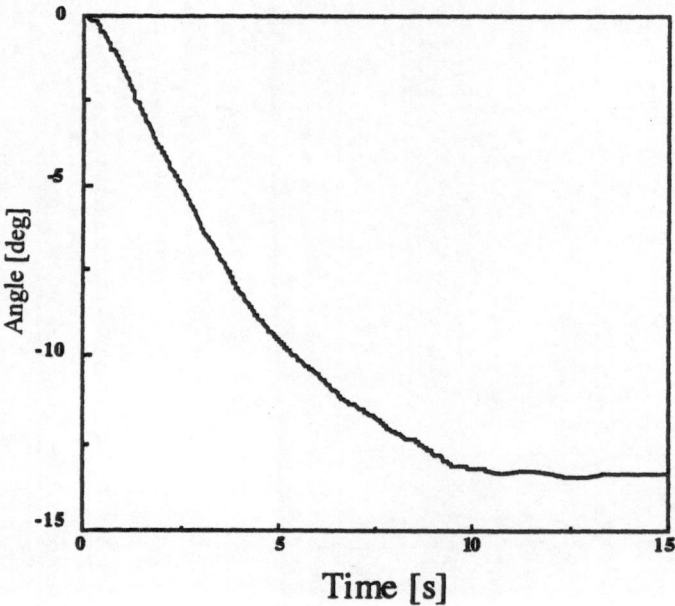

Figure 9: Euler angle θ_6 (Arm2)

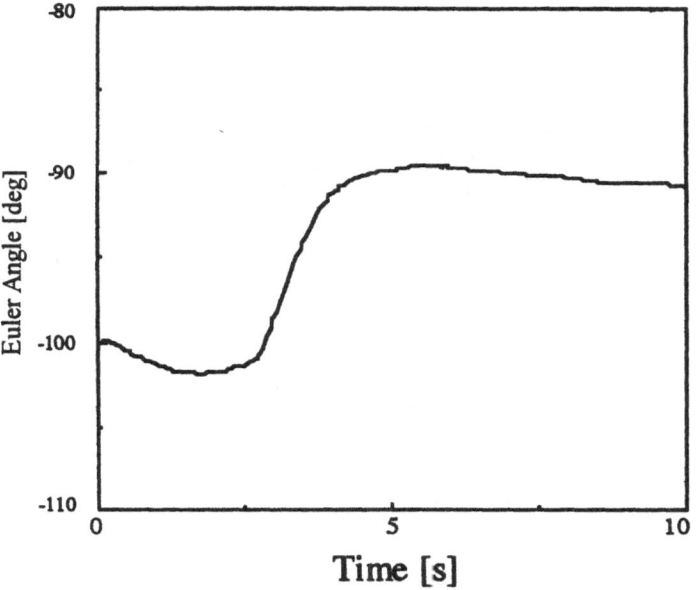

Figure 10: Euler angle γ_1 (Arm1)

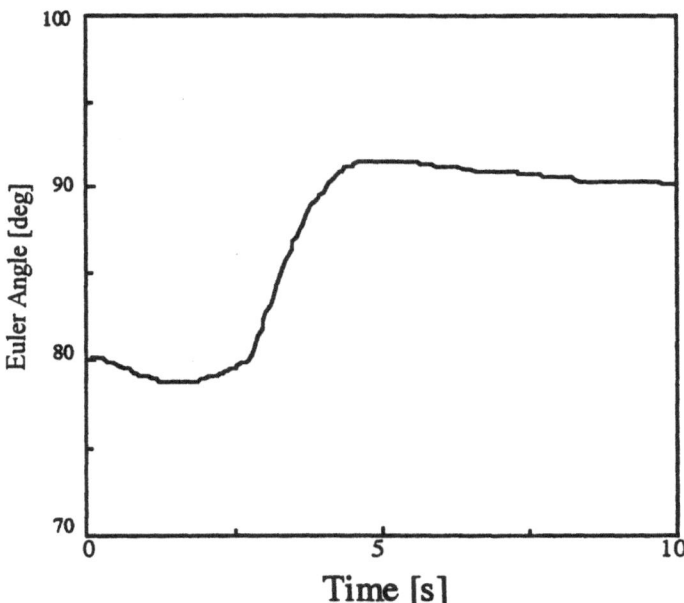

Figure 11: Euler angle γ_2 (Arm2)

174

6. Conclusion

In this paper, we modeled a flexible metal sheet based on the finite element when moments are exerted on the metal sheet. To control each manipulator around the representative point of a metal sheet, we proposed the control algorithm to realize the desired compliance of a metal sheet. The algorithm is applicable to both bending a metal sheet and pressing it along the direction of high rigidity. Experimental results illustrated the proposed control algorithm.

References

1.   K.Kosuge, H.Yoshida, T.Fukuda et al.: *"Unified Control for Dynamic Cooperative Manipulation"*, In: IEEE Int. Conf. on Intelligence Robots and Systems, pp.1042-1048, 1994.

2.   K.Kosuge, H.Yoshida, T.Fukuda et al., : *"Manipulation of a Flexible Object"*, In: IEEE Int. Conf. On Robotics and Automation, pp.318-323, 1995.

3.   K.Kosuge, S.Hashimoto, and H.Yoshida, :*"Human-Robots Collabolation System for Flexible Object Handling"*, In: IEEE Int. Conf. on Intelligence Robots and Systems, pp.1841-1846, 1998.

4.   Y.F.Zheng, Run Pei, and Chichyang, :*"Strategic for Automatic Assembly of Deformable Objects"*, In: IEEE Int. Conf. on Intelligence Robots and Systems, pp.2598-2603, 1996.

5.   Dong Sun, Yunhi Liu, and James K.Mills, :*"Cooperative Control of a two Manipulator System Handling a General Flexible Objects"*, In: IEEE Int. Conf. on Intelligence Robots and Systems, pp.5-10, 1997.

6.   Warner Kraus Jr, and Breman J McCarragher, :*"Hybrid Position Force Coordination for Dual-Arm Manipulation of Flexible Materials"*, In: IEEE Int. Conf. on Intelligence Robots and Systems, pp.202-207, 1997.

Section 4.2

A Manipulated Deformable Object as an Underactuated Mechanical System

H. G. Tanner and K. J. Kyriakopoulos

Abstract. Deformable objects under manipulation can be modeled using finite elements. The resulting model is in fact an underactuated mechanical system. The consequences of any type of constraints revealed by the modeling procedure is explored. The study of deformable object models within the framework of underactuated mechanical systems indicated the existence of second order nonholonomic constraints. For the identification of this kind of constraints, the authors have developed computationally efficient and significantly simpler mathematical tools. Their methodology is illustrated and tested by an example.

1. Introduction

In the study of robotics, the object that was being manipulated was traditionally considered rigid. This assumption on the object's nature simplified significantly the problem. The approach was justified from nearly every point of view. It allowed researchers to focus on the new system being developed, namely the robot, and assumed an ideal environment: no obstacles present, clean and absolutely known environment, rigid objects.

As the problems involved find their solution with time and great effort by the robotics community, all of these assumptions are gradually relieved. Obstacle avoidance has been considered, adaptation methods explored, sensor information taken into account. It is about time we did something about the assumption concerning the object being rigid.

In what follows we shall discuss some properties of a model obtained for deformable objects being manipulated by robotic arms. The motivation for investigating deformable materials came from the study of a multiple mobile manipulator system which handles a deformable object [1]. When focusing on the object issue, we first had to choose an appropriate model to describe its behavior.

Previous approaches to deformable object handling focused mainly on formulating general continuous dynamic equations for the object. The aim was to construct an appropriate model to enhance computation or allow for the application of certain control strategies. Sun et al. [2] followed Terzopoulos' [3] hybrid approach to deformable objects. This approach originated from the study of Computer Aided Design and computer graphics. It is characterized by the decomposition of the deformable object to a reference component and a deformation component. The former represents the original shape of the object and the latter the change in its shape as a result of the applied load. The position of any point in the body can be then determined by a superposition of the two components.

Kosuge et al. [4] used finite elements. Loading conditions included only bending of a sheet metal and examined the problem of controlling the static deformation of the plate when handled by a dual manipulation system. The static assumption simplifies considerably the dynamics of the object.

Wu et al. [5] pointed out that a flexible object is actually a distributed parameter system and approximated it by considering a lumped parameter model. They also assumed, quite reasonably, that the object is not very soft which means that it undergoes only small deflections during its interaction with the robot. A case of a flexible metal plate is considered. To construct the lumped parameter system they exploited the geometry of the system of the robot grasping the metal plate. It is not clear, however, how this can be extended to the case of other objects with arbitrary shape.

Yukawa et al. [6] investigated a vibrating flexible object and modeled it using model reduction theory. The aim was to realize position control while suppressing the vibration of the object. Vibration is assumed to take place in a two dimensional space. The modeling approach for the object begins with the distributed parameter model and ends up with a finite dimensional model.

In our approach to modeling deformable objects we decided to use finite elements based on the elastodynamic equations, motivated mainly by an engineering attitude towards the problem in hand. This step has a decisive impact on the nature of the problem we would then have to face. A question arises concerning the structural properties of the model thus obtained; how does that behave from the control point of view; what are the limitations and what can one hope to achieve.

Talking about achievement, we would better determine what needs to be done first. In our study of the multiple mobile manipulator system we placed emphasis on the deformable object, since this defines at a great degree the whole system's specifications and requirements. Everything has to be done with an eye on the object: should it be glass it must not break, should it be paper, cloth or leather it must not tear apart, should it be wood it must not fracture, should it be a metal plate it must not bend too much, and so on. We therefore needed to plan the motion of the object so that it can be safely been transferred from one point to another, without placing excessive load on the robots which could tip over or reach a singular position in their effort to accomplish their mission. That is why we had to see what the model we chose for the object could do.

First, we identified an underactuated system [7]. Finite element procedures result to a model which has a few points, or nodes, where load is directly exerted and a number of intermediate points which are more or less influenced by that load. Seen from the control perspective, the nodes where load is applied are directly actuated while the rest are statically and dynamically coupled to them. Therefore, there is a number of degrees of freedom which are actuated and others which are not. This is roughly the idea of an underactuated system: underactuated systems have fewer control inputs than degrees of freedom.

Viewing the manipulated object as an underactuated system we can considerably broaden the notion of a "deformable object" to the case of systems which are not continuous: chains, structures with passive joints, flexible links, rolling contacts, etc. This allows a generalization of our approach to object handling and puts a great variety of systems in a unified framework of study.

In the analysis of underactuated systems, it will soon become clear that a part of their dynamics, (which others prefer to call the zero dynamics), can be thought of as a set of dynamic constraints. In this context we refer to these equations as intrinsic constraints, since they stem directly from the nature of the system and are present regardless of the assumptions one may make for the manipulated object. Apart from these, practice can enforce other limitations such as the avoidance of fracture or excessive loading which can result to plastic (non-elastic) deformation. These external limitations can relate to material strength limitations, conditions for obstacle avoidance during the motion of the object, and so forth. Normally, the external constraints can be expressed in the form of a set of inequalities.

1.1 The Underactuated System

Strictly speaking, a deformable object has an infinite number of degrees of freedom. This is because every point in the body can be considered as an individual degree of freedom. The deeper one goes in a microscopic scale, the bigger the number grows. Although mathematicians would feel quite comfortable with the notion of infinity in the problem, this would probably distress most engineers who are assigned to the task of proposing a realistic solution.

Engineers would therefore try to decrease this number to a finite one and make a tradeoff between computational complexity and accuracy of solution. In the process of this simplification in order to deal with the problem using existing well known and approved tools, a natural approach is to "discretize" the object: consider it as a system of identical interconnected material elements that each contributes by a small portion to the system's overall behavior. This is actually the idea underlying the finite elements approach.

The extend of discretization depends on many factors. One is the individual characteristics of the material: objects that are fairly rigid do not require much discretization; flexible objects need many elements to sufficiently describe their shape. Another factor is loading conditions: for the same object, if a large load is to be applied the discretization grid should be denser since each element would be called to undertake a larger displacement. If the measure of this displacement is excessive, then the assumptions on which the approximation is based could cease to hold and the results may not be valid any more. For a given material and approximately known loading conditions, a pretty good discretization grid can be constructed.

The finite element method yields a dynamic model in which some degrees of freedom are driven directly as a result of external load and the rest comply with the displacement of the former. The degrees of freedom that are directly controlled correspond to the nodes of the grid at the location where the manipulators grasp the object. All other nodes are displaced in accordance to the loading conditions imposed by the driven nodes. The whole system can be therefore classified as underactuated [8] since the number of degrees of freedom being directly controlled is less than the total number of degrees of freedom in the system.

The class of underactuated mechanical systems is very broad. It includes systems with passive joints, flexible link constructions, chain mechanisms, mobile robots, and many others. These systems are peculiar in the sense that their Lagrangian dynamics may contain undesirable properties such as non-minimum phase zero dynamics, nonholonomic constraints, etc. Viewing the manipulated object as an underactuated system allows for the investigation of new manipulation tasks, including systems of rigid bodies or combinations of rigid and deformable objects. The object being handled can now be a truss, a frame, a plate, a chain or an arbitrary shaped three dimensional body and the material considered can be practically anything that humans themselves can manipulate.

In Lagrangian dynamics the equations of motion for a mechanical system with n degrees of freedom can be derived as

$$\frac{d}{dt}\left(\frac{\partial L}{\partial \dot{q}_i}\right) - \frac{\partial L}{\partial q_i} = \mathbf{F}_i \quad i = 1, \ldots, n,$$

where q are the generalized coordinates chosen for the system:

$$\mathbf{q} = \begin{pmatrix} q_1 & \cdots & q_n \end{pmatrix}^T \in \mathbf{Q} \subset \mathfrak{R}^n$$

It is well known that these equations can take the matrix equation form:

$$\mathbf{M(q)\ddot{q}} + \mathbf{C(q,\dot{q})\dot{q}} + \mathbf{K(q)} = \mathbf{B(q)u} \qquad (1)$$

where $\mathbf{M}$ is the inertia matrix, which is symmetric and positive definite, $\mathbf{C}$ is a matrix related to Coriolis and centrifugal forces, $\mathbf{K}$ is the vector formed by the elastic and gravitational terms and $\mathbf{u}$ is the input to the system. For the case of underactuated systems, $\mathbf{u} \in \mathfrak{R}^m$, with $m < n$.

The different dimension between the space of control inputs and the space of generalized coordinates motivates the partition of the latter into an actuated and an unactuated subspace. Without loss of generality we can rearrange the terms in the q vector as follows:

$$\mathbf{q}^T = \begin{pmatrix} \mathbf{q}_1^T & \mathbf{q}_2^T \end{pmatrix}$$

where $\mathbf{q}_1 \in \mathbf{Q}_1 \subset \mathfrak{R}^{n-m}$ corresponds to the unactuated coordinates and $\mathbf{q}_2 \in \mathbf{Q}_2 \subset \mathfrak{R}^m$ to the actuated or controlled coordinates. In the light of the above partition, equation (1) can be written [8]

$$\mathbf{m}_{11}\ddot{\mathbf{q}}_1 + \mathbf{m}_{12}\ddot{\mathbf{q}}_2 + \mathbf{c}_1(\mathbf{q},\dot{\mathbf{q}}) + \mathbf{k}_1(\mathbf{q}) = \mathbf{0} \qquad (2a)$$
$$\mathbf{m}_{21}\ddot{\mathbf{q}}_1 + \mathbf{m}_{22}\ddot{\mathbf{q}}_2 + \mathbf{c}_2(\mathbf{q},\dot{\mathbf{q}}) + \mathbf{k}_2(\mathbf{q}) = \mathbf{b}(\mathbf{q})\mathbf{u} \qquad (2b)$$

where the matrix $\mathbf{b}(\mathbf{q}) \in \mathfrak{R}^{m \times m}$ is assumed nonsingular.

Equation (2a) is a set of n-m second order uncontrolled differential equations which can be considered as dynamic constraints. Some prefer to interpret these equations as the zero dynamics, however this is not formally correct since no output has been specified for this system. In this context we will name these equations "intrinsic constraints" to distinguish them from any external imposed constraints that can relate to material strength limitations or obstacle avoidance requirements.

A natural question to ask is what kind of constraints these are. Can these differential equations be integrated once? If so, then one has a set of expressions relating the generalized velocities with the generalized coordinates, i. e. there will be no acceleration terms present. Do these resulting equations permit a second final integration? Should such integration be permitted, the result would be a set of algebraic equations relating the generalized coordinates.

In this final case, things are rather simple: The algebraic equations thus obtained define an *n-m* hypersurface on which the system must live. This is the case of holonomic constraints. For an object being manipulated, this means that the robots have no control whatsoever over *n-m* degrees of freedom of the system. These degrees are completely specified by the remaining *m*. Therefore, they cease to be degrees of freedom: they are just some fully dependent variables that can be equally ignored in the dynamic model. This could indicate a rigid behavior exhibited by a part of the object. In fact, such a situation is improbable, since even "rigid" material exhibits some sort of elastic behavior, at least in theory. It may arise however if during the initial modeling procedure some physical constraints have been overlooked, and the model comes later on to bring them into evidence. This implies that there is a way to improve the model and reduce the complexity of the problem.

If the equations can be integrated only once, one ends up with a set of first order nonholonomic constraints. There are only generalized coordinates and their first order time derivatives present. Now things are more complicated: these relations imply that there are limitations in the space of velocities that do not influence directly the space of positions. The influence of these velocity limitations is quite subtle. They impose significant limitations in control. In simple terms they imply that one may be able to steer the system from one place to another but the path it must follow is not necessarily the shorter in some respect. A simple example is parking a car: there is a steering sequence you have to follow – one cannot simply turn the car in any desirable direction. If this is the case, then for the task of planning trajectories for the object one has to resort to special and heavy mathematical tools.

The last case is when the equations cannot be integrated at all. This implies the presence of second order nonholonomic constraints. The deeper impact of such constraints has not been very well understood yet. Such systems could be hard to control, as in the case of first order nonholonomic constraints, but sometimes the systems can even be controlled with linear time-invariant controllers [8]. This is typical for systems that have an elastic/gravity (potential) term. The underlying characteristics of second order nonholonomic constraints will be clarified as research continues.

It is therefore important to investigate the type of the constraints imposed on the system, since these could determine to a certain degree the controllability properties of the system, or imply ways in which the model can be improved.

2. Collocated Linearization

Certain forms of system description enhance analysis. A notable one is the normal form, which will be used in subsequent analysis. This form reveals a structure that is particularly useful in identifying the type of constraints imposed on the system. Many of the previous results concerning nonholonomic constraints have used this type of description.

The normal form is obtained by a standard technique of feedback linearization. The system (2a-2b) can always be partially feedback linearized with respect to the actuated degrees of freedom. This is an important property of such systems [8]. The process of partial feedback linearization for this class of systems is also known as collocated linearization.

It can easily be understood that the matrix $\mathbf{m}_{11}$ in equation (2a) is square and nonsingular. The latter is guaranteed by the positive definiteness of the inertia matrix $\mathbf{M}$. If (2a) is solved for $\ddot{\mathbf{q}}_1$, we obtain

$$\ddot{\mathbf{q}}_1 = -\mathbf{m}_{11}^{-1}\left[\mathbf{m}_{12}\ddot{\mathbf{q}}_2 + \mathbf{c}_1(\mathbf{q},\dot{\mathbf{q}}) + \mathbf{k}_1(\mathbf{q})\right]$$

This can be used to substitute $\ddot{\mathbf{q}}_1$ in (2b) so that it becomes:

$$\overline{\mathbf{m}}(\mathbf{q})\ddot{\mathbf{q}}_2 + \overline{\mathbf{c}}(\mathbf{q},\dot{\mathbf{q}}) + \overline{\mathbf{k}}(\mathbf{q}) = \mathbf{b}(\mathbf{q})\mathbf{u}$$

where

$$\overline{\mathbf{m}}(\mathbf{q}) = \mathbf{m}_{22}(\mathbf{q}) - \mathbf{m}_{21}(\mathbf{q})\mathbf{m}_{11}^{-1}(\mathbf{q})\mathbf{m}_{12}(\mathbf{q})$$

$$\overline{\mathbf{c}}(\mathbf{q},\dot{\mathbf{q}}) = \mathbf{c}_2(\mathbf{q},\dot{\mathbf{q}}) - \mathbf{m}_{21}(\mathbf{q})\mathbf{m}_{11}^{-1}(\mathbf{q})\mathbf{c}_1(\mathbf{q},\dot{\mathbf{q}})$$

$$\overline{\mathbf{k}}(\mathbf{q}) = \mathbf{k}_2(\mathbf{q}) - \mathbf{m}_{21}(\mathbf{q})\mathbf{m}_{11}^{-1}(\mathbf{q})\mathbf{k}_1(\mathbf{q})$$

Using now the linearizing feedback

$$\mathbf{u} = \mathbf{b}(\mathbf{q})^{-1}\left[\overline{\mathbf{m}}(\mathbf{q})\mathbf{v} + \overline{\mathbf{c}}(\mathbf{q},\dot{\mathbf{q}}) + \overline{\mathbf{k}}(\mathbf{q})\right] \tag{3}$$

where v is the new control vector, the system (2a – 2b) can take the form

$$\ddot{\mathbf{q}}_2 = \mathbf{v} \tag{4a}$$

$$\tag{4b}$$

$$\ddot{q}_1 = J(q)\ddot{q}_2 + R(q, \dot{q})$$

where the terms are defined as follows

$$J(q) = -m_{11}^{-1}(q)\,m_{12}(q)$$

$$R(q, \dot{q}) = -m_{11}^{-1}(q)\,c_1(q, \dot{q}) - m_{11}^{-1}(q)k_1(q)$$

2.2 State Space Description

The state space equations of the underactuated system can be easily obtained by equations (4) by setting:

$$x_1 = q_1 \in \Re^{n-m}, \qquad x_2 = q_2 \in \Re^m, \quad x_3 = \dot{q}_1, \qquad x_4 = \dot{q}_2$$

This way, equations (4) can take the form:

$$\dot{x}_1 = x_3 \tag{5a}$$

$$\dot{x}_2 = x_4 \tag{5b}$$

$$\dot{x}_3 = J(x_1, x_2)\,v + R(x_1, x_2, x_3, x_4) \tag{5c}$$

$$\dot{x}_4 = v \tag{5d}$$

with configuration vector $x = \begin{bmatrix} x_1 & x_2 \end{bmatrix}^T \in M \subset \Re^n$. The above equations have an unactuated linear part (5a – 5b) and an actuated nonlinear part (5c – 5d). They will be the starting point for the analysis of the constraint equations and the stabilization properties that will follow. We chose to rename the configuration variables because the partition of the original configuration space and the rearrangement of variables can easily cause confusion.

3. Constraint Classification

3.3 Preliminaries

In the sequel we use some mathematical concepts from the field of differential geometry. For the reader who is not familiar we these terms we will attempt a short and informal introduction. Those familiar with the terms may skip this section. The definitions given informally below are by no means complete nor acceptable mathematically and they serve only to allow a non-specialist to follow intuitively our approach. The interested reader can refer to [9].

A *manifold* is a locally Euclidean space. Locally means that one cannot use the same constructions (not even the same coordinate system) to move from a point to any other point. A sphere is a manifold which is only locally Euclidean. We consider the earth's surface as a plane because in our scale it seems so, and practically it is a excellent approximation. Everyone however has a clear picture of the spherical shape of Earth and does not expect to join the north and the south pole with a straight line without that line crossing the surface. The planar approximation can only hold locally.

A *tangent vector* is a vector tangent to the manifold's surface attached to a point on the manifold. The tangent vector is always referred to in connection to the point at which it is attached. All tangent vectors at a point form the *tangent space* at that point. A *vector field* is a mapping that assigns to each point on the manifold, a tangent vector on that manifold. The velocity of a particle moving on the manifold is a vector field. Vector fields are closely related to differentiation. That is why we use the symbols $\partial / \partial x$ to denote the base vectors in the tangent space. The right hand side of the state equations of the system can define a set of vector fields $\mathbf{g}_i(\mathbf{x})$:

$$\dot{\mathbf{x}} = \mathbf{g}_1(\mathbf{x}) v_1 + \cdots + \mathbf{g}_p(\mathbf{x}) v_p$$

Having one or more vector fields on a manifold you can assign at each point one or more tangent vectors. These tangent vector may span a tangent subspace at this point. The collection of all tangent subspaces generated by the tangent vectors of the vector fields forms a *distribution*.

Several operations can be defined on vector fields. One of them is the *Lie bracket* which resembles an outer product operation. The product of the Lie bracket operation between two vector fields is another vector field. Under the Lie bracket operations the vector fields can form an algebra, namely the *Lie algebra*. With every distribution spanned by some vector fields an algebra is associated. Studying the mathematical properties of this algebra one can draw significant conclusions concerning the controllability properties of a system the vector fields of which generate the associated distribution.

3.4 Definitions and Constraint Identification

Nonholonomic constraints can be identified with the use of some tools from differential geometry. Consider equation (5). We state the following definition, adopted from [10]

Definition 1 [10]: *Consider the system (5) and define the following vector fields:*

$$\tau_0 = \sum_{j=1}^{m} \dot{x}_{2,j} \frac{\partial}{\partial \dot{x}_{2,j}} + \sum_{k-1}^{n-m} \left(\dot{x}_{1,k} \frac{\partial}{\partial x_{1,k}} + R_k \frac{\partial}{\partial \dot{x}_{1,k}} \right) + \frac{\partial}{\partial t},$$

$$\tau_j = \frac{\partial}{\partial \dot{x}_{2,j}} + \sum_{i=1}^{n-m} J_{ij} \frac{\partial}{\partial \dot{x}_{1,i}}, \qquad j = 1, \ldots, m$$

and let $\Delta = span\{\tau_0, \tau_j\}$ *be the distribution generated by them. Consider the accessibility algebra* $\tilde{C}$ *of the distribution* Δ, *i. e. the smallest subalgebra that contains the vector fields* $\{\tau_0, \tau_j\}$. *Let* $\tilde{C}$ *be the accessibility distribution generated by the accessibility algebra. If* $\dim \tilde{C}(\mathbf{x}, t) = 2n + 1 \quad \forall (\mathbf{x}, t) \in \mathbf{M} \times \mathfrak{R}$, *then the system (5) is called completely second order nonholonomic.*

When the condition for the dimension of the accessibility algebra is satisfied for the underactuated system (5), then it possesses second order nonholonomic constraints. This implies that the constraint equations (2b) cannot be integrated to produce a distribution for the system. One should carefully distinguish between the integration with respect to time and the integration we are discussing. The latter refers to Frobenius integration, i. e. finding smooth functions λ_i that solve a partial differential equation $\dfrac{\partial \lambda_i}{\partial [\mathbf{x} \quad \dot{\mathbf{x}}]} \begin{bmatrix} \tau_0 & \tau_j \end{bmatrix} \equiv \mathbf{0}$. If such an integration is possible then it means that the dynamic equations restrict the system to develop on an $n+m$ dimensional distribution. If on the other hand, the constraints are second order nonholonomic, the dimension of the state space remains $2n$. Whether these constraints influence decisively the controllability properties of the system is another issue which will be discussed later.

Suppose now that the constraints are not second order nonholonomic and are integrated once. The resulting equations will now relate the generalized coordinates with their first time derivatives. The resulting equations can then be expressed in the form:

$$A(\mathbf{q})\dot{\mathbf{q}} = \mathbf{0}, \qquad A \in \mathfrak{R}^{(n-m) \times n} \tag{6}$$

The annihilators of the rows of matrix $A(\mathbf{q})$ form another matrix, $S(\mathbf{q}) \in \mathfrak{R}^{m \times n}$, for which

$$A(\mathbf{q})S(\mathbf{q}) \equiv \mathbf{0}$$

The existence of $S(\mathbf{q})$ implies a relationship of the form [11]

$$\dot{\mathbf{q}} = S^T(\mathbf{q})\varsigma, \qquad \varsigma \in \mathfrak{R}^m \tag{7}$$

The columns of $S^T(\mathbf{q})$ are vector fields which define a distribution. Consider the accessibility algebra that is generated by those vector fields (the smaller subalgebra that contains those vector fields). If the accessibility algebra has a dimension equal to n, then the system is completely *first order nonholonomic* by Frobenius theorem. This means that the system is restricted to evolve on a tangent bundle of dimension $n+m$ which can be readily shown with the standard analysis that will follow. Nonholonomic constraints allow the system to reach any desired position, however they locally limit the directions on which the system can move. The limitation does not apply to configuration variables but rather to velocities: at a specific configuration the system may not be able to develop velocity at certain directions. The number $n+m$ can be thought of as the new dimension of the state space.

After differentiation, equation (7) becomes

$$\ddot{\mathbf{q}} = S^T(\mathbf{q})\dot{\varsigma} + \dot{S}(\mathbf{q})\varsigma$$

This can be substituted in (1) to yield

$$\mathbf{M(q)}\,(\mathbf{S}^T(\mathbf{q})\dot{\mathbf{\varsigma}} + \dot{\mathbf{S}}(\mathbf{q})\mathbf{\varsigma}) + \mathbf{C(q,\dot{q})\dot{q}} + \mathbf{K(q)} = \mathbf{B(q)u}$$

and multiplying by $\mathbf{S(q)}$ from the right

$$\mathbf{S\,M\,S}^T\dot{\mathbf{\varsigma}} + \left\{\mathbf{S\,M\,\dot{S}\varsigma} + \mathbf{S\,C(q,S}^T\mathbf{\varsigma})\,\mathbf{S}^T\mathbf{\varsigma}\right\} + \mathbf{S\,K} = \mathbf{S\,B\,u} \tag{8}$$

where the dependence on $\mathbf{q}$ has been dropped.

Equation (8) is now m dimensional and give the reduced dynamic model of the system (1). After some algebraic manipulation and rearranging of terms, equation (8) can be written as

$$\dot{\mathbf{\varsigma}} = \mathbf{D(q)} + \mathbf{G(q,\varsigma)\,u}$$

and form the new reduced order model of the system

$$\dot{\mathbf{q}} = \mathbf{S}^T\mathbf{\varsigma} \tag{9}$$

$$\dot{\mathbf{\varsigma}} = \mathbf{D(q)} + \mathbf{G(q,\varsigma)\,u}$$

which after rearranging the terms can be brought to the state space form (5).

Suppose now that equations (6) can still be integrated. The result of this integration is a set of $n\text{-}m$ *holonomic* equations, prescribing $n\text{-}m$ degrees of freedom in terms of the remaining m:

$$\mathbf{g(q_1,q_2)} = \mathbf{0}$$

The implicit function theorem provides the necessary conditions under which the above equation can be solved for $\mathbf{q}_1$ to yield:

$$\mathbf{q}_1 = \mathbf{h(q_2)}$$

The above relation can be used to eliminate $\mathbf{q}_1$ from equation (2b) so as to obtain a reduced order model.

$$\hat{\mathbf{M}}(\mathbf{q}_2)\ddot{\mathbf{q}}_2 + \hat{\mathbf{C}}(\mathbf{q}_2,\dot{\mathbf{q}}_2)\dot{\mathbf{q}}_2 + \hat{\mathbf{K}}(\mathbf{q}_2) = \hat{\mathbf{B}}(\mathbf{q}_2)\mathbf{u}, \quad \mathbf{q}_2 \in \mathbf{Q}_2 \subset \mathfrak{R}^m$$

Another case is when the dynamic constraints that form equation (2a) are a collection of second order nonholonomic, first order nonholonomic and even holonomic. In this occasion, the procedure outlined for each case should be followed for the part that falls within each category. This could be quite troublesome but it is the only way to clear out the scene and obtain a consistent, minimum order model.

4. Controllability Issues

When k dynamic constraints are holonomic, then the system motion is restricted to an $n\text{-}k$ dimensional manifold. It cannot reach any position outside this manifold. Points that do not belong to that manifold are simply unreachable.

In the case of completely first order nonholonomic constraints, the configuration space is not confined. The system is known to be accessible. Due, however, to the drift term $\mathbf{D(q)}$ in equation (9), accessibility for the system does not imply controllability. This does not mean that the system is not controllable; it simply means that there is no definite clue that it is. For nonholonomic systems with drift, there is no available general necessary and sufficient result for establishing complete controllability [12]. One has to resort to other forms of controllability, such as strong accessibility and small-time-local controllability. For the latter, there exist only sufficient conditions, but once it has been established one can use the manifold of equilibrium points of the drift vector field to reach an arbitrary small neighborhood of the desired configuration. A difficult point is that first order nonholonomic systems are not stabilizable via continuous time-invariant state feedback. In summary, this is a situation that one should wish to avoid.

If however the system (5) is proved to be second-order nonholonomic, it has been proved [10] that it is automatically strongly accessible. Moreover, there is even a chance for smooth feedback stabilization, provided that a sufficient condition for non-existence of a smooth stabilizing control law is not satisfied:

Theorem 1 [10]: *Assume that* $R_i(\mathbf{x},\mathbf{0}) = 0, \quad \forall \mathbf{x} \in \mathbf{M}$, *for i=1, ..., n-m.*

Let $n - m \geq 1$ *and let* $(\mathbf{x}^e, \mathbf{0})$ *denote an equilibrium solution. Then the second order nonholonomic system (5) is not asymptotically stabilizable to* $(\mathbf{x}^e, \mathbf{0})$, *using time-invariant continuous (static or dynamic) state feedback law.*

If on the other hand $\forall i, \exists \mathbf{x} \in \mathbf{M} \mid R_i(\mathbf{x},0) \neq 0$, then the system could perhaps be stabilizable by continuous control law.

4.5 The Finite Element Model

Lets return now to the finite element model we have developed for the object. The finite element analysis results to a dynamic model which is linear [13]:

$$\mathbf{M}\ddot{\mathbf{q}} + \mathbf{C}\dot{\mathbf{q}} + \mathbf{K}\mathbf{q} = \mathbf{B}\mathbf{u} \qquad (10)$$

Moreover, the characteristic matrices $\mathbf{M}$, $\mathbf{C}$, $\mathbf{K}$ and $\mathbf{B}$ are independent of the node coordinates. The finite element method provides a meaningful way of linearizing the original dynamic equations of the deformable object. Under specific conditions for selecting the interpolation functions within the element, the linear model can be proved to be equivalent to the initial nonlinear differential equation, so that none of the information contained in the original equations is lost in the process and all modes are represented by the approximated model.

The state equations derived from (10) have the form

$$\ddot{\mathbf{q}}_2 = \mathbf{v}$$
$$\ddot{\mathbf{q}}_1 = \mathbf{J}\ddot{\mathbf{q}}_2 + \mathbf{R}(\mathbf{q},\dot{\mathbf{q}})$$

The state equations can be formed as

$$\dot{\mathbf{x}}_1 = \mathbf{x}_3$$
$$\dot{\mathbf{x}}_2 = \mathbf{x}_4$$
$$\dot{\mathbf{x}}_3 = \mathbf{J}\,\mathbf{v} + \mathbf{R}(\mathbf{x}_1, \mathbf{x}_2, \mathbf{x}_3, \mathbf{x}_4) \qquad (11)$$
$$\dot{\mathbf{x}}_4 = \mathbf{v}$$

We give the following Lemma:

Lemma 1 : *For the system (11) it holds:*

1. $\left[\tau_k, ad^r_{\tau_0}\tau_j\right] = 0, \qquad \forall r \geq 0, \qquad k, j \in \{1, \ldots, m\}$

2. $ad^{1+r}_{\tau_0}\tau_j = (-1)^{r-1}\begin{bmatrix} A_r \cdot \left[\mathbf{J}^T_{\ j} \quad \mathbf{e}^T_{\ j}\right]^T \\ \mathbf{0}_{m\times 1} \\ B_r \cdot \left[\mathbf{J}^T_{\ j} \quad \mathbf{e}^T_{\ j}\right]^T \\ \mathbf{0}_{(m+1)\times 1} \end{bmatrix}$ *for $r > 0$, $A_r, B_r \in \mathfrak{R}^{(n-m)\times 1}$, $\mathbf{J}_j$*

   the jth column of $\mathbf{J}$ and $\mathbf{e}_j$ is the jth base vector of $\mathfrak{R}^m$

3. *The vectors that form the vector field $ad^{1+r}_{\tau_0}\tau_j$ can be expressed as follows:*

$$B_r = \frac{\partial \mathbf{R}}{\partial \mathbf{x}_1} A_{r-1} + \frac{\partial \mathbf{R}}{\partial \dot{\mathbf{x}}_1} B_{r-1} \qquad B_1 = \frac{\partial \mathbf{R}}{\partial \mathbf{x}} + \frac{\partial \mathbf{R}}{\partial \dot{\mathbf{x}}_1} \cdot \frac{\partial \mathbf{R}}{\partial \dot{\mathbf{x}}}$$

$$A_r = B_{r-1} \qquad \qquad \qquad A_1 = \frac{\partial \mathbf{R}}{\partial \dot{\mathbf{x}}}$$

with

Proof:

1. Define $\overline{\mathbf{x}} = [\mathbf{x} \quad \dot{\mathbf{x}} \quad t]$. For the Lie bracket it is $[\tau_k, ad^r_{\tau_0}\tau_j] = [\tau_k, [\tau_0, ad^{r-1}_{\tau_0}]]$. For $r=0$ we have $[\tau_k, \tau_j] = \frac{\partial \tau_j}{\partial \overline{\mathbf{x}}}\tau_k - \frac{\partial \tau_k}{\partial \overline{\mathbf{x}}}\tau_j$. Recalling Definition 1, and calculating τ_j for the

system (11), we can see that $\tau_j = \begin{bmatrix} \mathbf{0}_{n\times 1} \\ \mathbf{J}_j \\ \mathbf{e}_j \\ 0 \end{bmatrix}$. However $\mathbf{J}$ is independent of $\overline{\mathbf{x}}$, so that

both $\frac{\partial \tau_j}{\partial \overline{\mathbf{x}}}$ and $\frac{\partial \tau_k}{\partial \overline{\mathbf{x}}}$ are zero. It is easy to see that any Lie bracket $[\tau_k, \mathbf{0}] = \mathbf{0}$.

2. and 3. It can be shown by induction: We first verify for $r = 1$. It is $ad_{\tau_0}^2 \tau_j = [\tau_0, [\tau_0, \tau_j]]$. The inner bracket is shown to be: $[\tau_0, \tau_j] = -\dfrac{\partial \tau_0}{\partial \overline{x}} \tau_j$, where

$$\frac{\partial \tau_0}{\partial \overline{x}} = \begin{bmatrix} \mathbf{0}_{n \times n} & \mathbf{I}_{n \times n} & \mathbf{0}_{n \times 1} \\ \left(\dfrac{\partial R}{\partial x}\right)_{(n-m) \times n} & \left(\dfrac{\partial R}{\partial \dot{x}}\right)_{(n-m) \times n} & \mathbf{0}_{(n-m) \times 1} \\ & \mathbf{0}_{(m+1) \times (2n+1)} & \end{bmatrix}.$$

Matrix $\mathbf{I}$ is the identity matrix. Now, for $r = 1$, $ad_{\tau_0}^2 \tau_j = \dfrac{\partial [\tau_0, \tau_j]}{\partial \overline{x}} \tau_0 - \dfrac{\partial \tau_0}{\partial \overline{x}} [\tau_0, \tau_j]$. By the structure of the finite element model (10), it can be seen that $\mathbf{R}$ does not contain quadratic terms in x, nor in $\dot{x}$. Therefore, $\dfrac{\partial \tau_0}{\partial \overline{x}}$ is independent of $\overline{x}$. On the other hand, τ_j is also constant since $\mathbf{J}$ is independent of $\overline{x}$.

Thus $\dfrac{\partial [\tau_0, \tau_j]}{\partial \overline{x}} = 0$ and $ad_{\tau_0}^2 \tau_j = -\dfrac{\partial \tau_0}{\partial \overline{x}} [\tau_0, \tau_j] = \left(\dfrac{\partial \tau_0}{\partial \overline{x}}\right)^2 \tau_j$. Calculating the square of this matrix one can verify that

$$\left(\frac{\partial \tau_0}{\partial \overline{x}}\right)^2 =$$

$$\begin{bmatrix} \left(\begin{array}{cc} \dfrac{\partial R}{\partial x_1} & \dfrac{\partial R}{\partial x_2} \\ \mathbf{0}_{m \times (n-m)} & \mathbf{0}_{m \times m} \end{array}\right)_{n \times n} & \left(\begin{array}{cc} \dfrac{\partial R}{\partial \dot{x}_1} & \dfrac{\partial R}{\partial \dot{x}_2} \\ \mathbf{0}_{m \times (n-m)} & \mathbf{0}_{m \times m} \end{array}\right)_{n \times n} & \mathbf{0}_{n \times 1} \\ \left(\begin{array}{cc} \left(\dfrac{\partial R}{\partial \dot{x}_1}\right)^2 & \dfrac{\partial R}{\partial \dot{x}_1} \cdot \dfrac{\partial R}{\partial x_2} \\ \mathbf{0}_{m \times (n-m)} & \mathbf{0}_{m \times m} \end{array}\right)_{n \times n} \left(\begin{array}{c} \dfrac{\partial R}{\partial x} \\ \mathbf{0}_{m \times n} \end{array}\right)_{n \times n} + \left(\begin{array}{cc} \left(\dfrac{\partial R}{\partial \dot{x}_1}\right)^2 & \dfrac{\partial R}{\partial \dot{x}_1} \cdot \dfrac{\partial R}{\partial \dot{x}_2} \\ \mathbf{0}_{m \times (n-m)} & \mathbf{0}_{m \times m} \end{array}\right)_{n \times n} & \mathbf{0}_{n \times 1} \\ \mathbf{0}_{1 \times n} & \mathbf{0}_{1 \times n} & 0 \end{bmatrix}$$

which multiplied by τ_j yields:

$$
ad_{\tau_0}^2 \tau_j = \begin{bmatrix} \dfrac{\partial \mathbf{R}}{\partial \dot{\mathbf{x}}} \cdot \begin{bmatrix} \mathbf{J}_j \\ \mathbf{e}_j \end{bmatrix} \\ \mathbf{0}_{m \times 1} \\ \left(\dfrac{\partial \mathbf{R}}{\partial \mathbf{x}} + \left(\left(\dfrac{\partial \mathbf{R}}{\partial \dot{\mathbf{x}}_1} \right)^2 \dfrac{\partial \mathbf{R}}{\partial \dot{\mathbf{x}}_1} \cdot \dfrac{\partial \mathbf{R}}{\partial \dot{\mathbf{x}}_2} \right) \right) \cdot \begin{bmatrix} \mathbf{J}_j \\ \mathbf{e}_j \end{bmatrix} \\ \mathbf{0}_{(m+1) \times 1} \end{bmatrix} \begin{bmatrix} \dfrac{\partial \mathbf{R}}{\partial \dot{\mathbf{x}}} \cdot \begin{bmatrix} \mathbf{J}_j \\ \mathbf{e}_j \end{bmatrix} \\ \mathbf{0}_{m \times 1} \\ \left(\dfrac{\partial \mathbf{R}}{\partial \mathbf{x}} + \dfrac{\partial \mathbf{R}}{\partial \dot{\mathbf{x}}_1} \cdot \left(\dfrac{\partial \mathbf{R}}{\partial \dot{\mathbf{x}}_1} \quad \dfrac{\partial \mathbf{R}}{\partial \dot{\mathbf{x}}_2} \right) \right) \cdot \begin{bmatrix} \mathbf{J}_j \\ \mathbf{e}_j \end{bmatrix} \\ \mathbf{0}_{(m+1) \times 1} \end{bmatrix}
$$

This can be written equivalently

$$
ad_{\tau_0}^2 \tau_j = \begin{bmatrix} \dfrac{\partial \mathbf{R}}{\partial \dot{\mathbf{x}}} \cdot \begin{bmatrix} \mathbf{J}_j \\ \mathbf{e}_j \end{bmatrix} \\ \mathbf{0}_{m \times 1} \\ \left(\dfrac{\partial \mathbf{R}}{\partial \mathbf{x}} + \dfrac{\partial \mathbf{R}}{\partial \dot{\mathbf{x}}_1} \cdot \dfrac{\partial \mathbf{R}}{\partial \dot{\mathbf{x}}} \right) \cdot \begin{bmatrix} \mathbf{J}_j \\ \mathbf{e}_j \end{bmatrix} \\ \mathbf{0}_{(m+1) \times 1} \end{bmatrix},
$$

which proves our result for $r = 1$.

Suggest now that it holds for $r = k$. Then for $r = k+1$,

$$
ad_{\tau_0}^{1+(k+1)} \tau_j = \frac{\partial\, ad_{\tau_0}^{1+k} \tau_j}{\partial \overline{\mathbf{x}}} \tau_0 - \frac{\partial \tau_0}{\partial \overline{\mathbf{x}}} ad_{\tau_0}^{1+k} \tau_j = -\frac{\partial \tau_0}{\partial \overline{\mathbf{x}}} ad_{\tau_0}^{1+k} \tau_j,
$$

equivalently,

$$
ad_{\tau_0}^{1+(k+1)} \tau_j = -\begin{bmatrix} \mathbf{0}_{n \times n} & \mathbf{I}_{n \times n} & \mathbf{0}_{n \times 1} \\ \left(\dfrac{\partial \mathbf{R}}{\partial \mathbf{x}} \right)_{(n-m) \times n} & \left(\dfrac{\partial \mathbf{R}}{\partial \dot{\mathbf{x}}} \right)_{(n-m) \times n} & \mathbf{0}_{(n-m) \times 1} \\ \multicolumn{3}{c}{\mathbf{0}_{(m+1) \times (2n+1)}} \end{bmatrix} \cdot (-1)^{k-1} \begin{bmatrix} A_k \cdot \begin{bmatrix} \mathbf{J}_j \\ \mathbf{e}_j \end{bmatrix} \\ \mathbf{0}_{m \times 1} \\ B_k \cdot \begin{bmatrix} \mathbf{J}_j \\ \mathbf{e}_j \end{bmatrix} \\ \mathbf{0}_{(m+1) \times 1} \end{bmatrix} =
$$

$$
= (-1)^{(k+1)-1} \begin{bmatrix} B_k \cdot \begin{bmatrix} \mathbf{J}_j \\ \mathbf{e}_j \end{bmatrix} \\ \mathbf{0}_{m \times 1} \\ \dfrac{\partial \mathbf{R}}{\partial \mathbf{x}_1} A_k \cdot \begin{bmatrix} \mathbf{J}_j \\ \mathbf{e}_j \end{bmatrix} + \dfrac{\partial \mathbf{R}}{\partial \dot{\mathbf{x}}_1} B_k \cdot \begin{bmatrix} \mathbf{J}_j \\ \mathbf{e}_j \end{bmatrix} \\ \mathbf{0}_{(m+1) \times 1} \end{bmatrix}
$$

which completes the proof.

We are now ready to present our main result:

Proposition 1: *Define the sequence of matrices:*

$$
\begin{bmatrix} A_1 \\ \mathbf{0}_{m \times 1} \\ B_1 \\ \mathbf{0}_{(m+1) \times 1} \end{bmatrix} \cdot \begin{bmatrix} J \\ \mathbf{I}_{m \times m} \end{bmatrix} \quad \cdots \quad \begin{bmatrix} A_r \\ \mathbf{0}_{m \times 1} \\ B_r \\ \mathbf{0}_{(m+1) \times 1} \end{bmatrix} \cdot \begin{bmatrix} J \\ \mathbf{I}_{m \times m} \end{bmatrix} \cdots
$$

where A_1, B_1, A_r, B_r are

$$
B_1 = \frac{\partial \mathbf{R}}{\partial \mathbf{x}} + \frac{\partial \mathbf{R}}{\partial \dot{\mathbf{x}}_1} \cdot \frac{\partial \mathbf{R}}{\partial \dot{\mathbf{x}}}
$$

$$
B_r = \frac{\partial \mathbf{R}}{\partial \mathbf{x}_1} A_{r-1} + \frac{\partial \mathbf{R}}{\partial \dot{\mathbf{x}}_1} B_{r-1}
$$

$$
A_1 = \frac{\partial \mathbf{R}}{\partial \dot{\mathbf{x}}}
$$

$$
A_r = B_{r-1}
$$

If for some r, the extended matrix

$$
\begin{bmatrix} \tau_0 & \tau_j & [\tau_0, \tau_j] & \begin{bmatrix} A_1 \\ \mathbf{0}_{m \times 1} \\ B_1 \\ \mathbf{0}_{(m+1) \times 1} \end{bmatrix} \cdot \begin{bmatrix} J \\ \mathbf{I}_{m \times m} \end{bmatrix} & \cdots & \begin{bmatrix} A_r \\ \mathbf{0}_{m \times 1} \\ B_r \\ \mathbf{0}_{(m+1) \times 1} \end{bmatrix} \cdot \begin{bmatrix} J \\ \mathbf{I}_{m \times m} \end{bmatrix} & \cdots \end{bmatrix}
$$

has rank 2n+1 then the system (11) is second order nonholonomic.

Proof: The proof follows from the Definition 1 and the previous Lemma. The series defined above is directly associated with the accessibility distribution. Indeed the vector fields $\tau_0, \tau_j, [\tau_0, \tau_j], ad_{\tau_0}^{1+r} \tau_j$ define one Phillip Hall basis for the accessibility distribution. It is easily shown that vector fields $\tau_0, \tau_j, [\tau_0, \tau_j]$ are linearly independent. This is obvious for τ_0, τ_j by their definition. On the other hand,

$$
[\tau_0, \tau_j] = - \begin{bmatrix} \mathbf{0}_{n \times n} & \mathbf{I}_{n \times n} & \mathbf{0}_{n \times 1} \\ \left(\frac{\partial \mathbf{R}}{\partial \mathbf{x}} \right)_{(n-m) \times n} & \left(\frac{\partial \mathbf{R}}{\partial \dot{\mathbf{x}}} \right)_{(n-m) \times n} & \mathbf{0}_{(n-m) \times 1} \\ \mathbf{0}_{(m+1) \times (2n+1)} \end{bmatrix} \cdot \begin{bmatrix} \mathbf{0}_{n \times 1} \\ \mathbf{J}_j \\ \mathbf{e}_j \\ 0 \end{bmatrix} = - \begin{bmatrix} \mathbf{J}_j \\ \mathbf{e}_j \\ * \\ 0 \end{bmatrix}
$$

As it can be seen by inspection, the *2m+1* vector fields $\tau_0, \tau_j, [\tau_0, \tau_j]$ are linearly independent and generate a *2m+1* dimensional distribution. Including more vector fields $ad_{\tau_0}^{r+1} \tau_j$ in the set, the dimension of the distribution grows. This sequence of distributions $G_i = G_{i-1} + [G_1, G_{i-1}]$ is a filtration. Each G_i is spanned by vector fields of the previous one plus some vector fields formed by taking *i-1* Lie brackets. If for some r, the rank of the extended matrix is 2n+1 is means that it contains 2n+1 linearly independent columns. The columns of the matrix, however, are exactly the vector fields that one would calculate for the P. Hall basis of the accessibility algebra. Choosing 2n+1 linearly independent columns one has 2n+1 independent vector fields that span the accessibility algebra of the system (11). By Definition 1, (11) is second order nonholonomic.

Generally, as it will be shown in the examples, systems derived by finite elements are usually second order nonholonomic. The process of determining the existence of second order nonholonomic constraints in systems with possibly hundred degrees of freedom through the conventional way of seeking for a Phillip Hall basis becomes awfully cumbersome. On the other hand, the algorithm just described provides an immediate way of investigation since one can easily automate the above procedure.

Being second order nonholonomic, the finite element model (11) is strongly accessible. One can easily verify however, that it does not satisfy the sufficient condition for small time local controllability, presented in [14]. However, this does not rule out the possibility that the system may be small time locally controllable. In fact, since the finite element model is linear, a simple matrix calculation can show whether that system is controllable. In that case the stabilizing feedback law is not only continuous but also linear! This comes in accordance to Theorem 1, since the

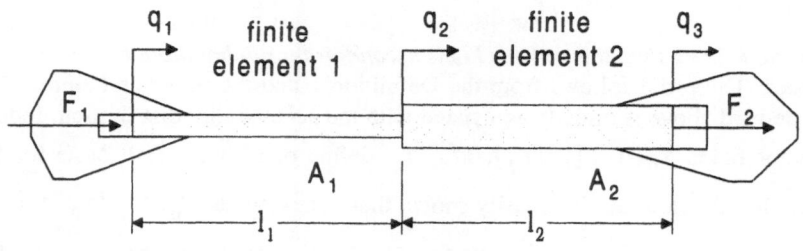

Figure 1: A deformable object under axial load

condition for nonexistence of a smooth feedback law, $\mathbf{R}(\mathbf{x}^e,\mathbf{0}) = \mathbf{0}, \quad \forall \mathbf{x}$, is not satisfied.

5. Example - Rod Under Axial Load

Consider a beam under axial load (Figure 1). The beam is divided into two finite elements. The system has three degrees of freedom, two of which are directly controlled. The element characteristic matrices are:

$$\mathbf{M} = \frac{\rho A \ell}{6} \begin{bmatrix} 2 & 1 \\ 1 & 2 \end{bmatrix}, \quad \mathbf{C} = \frac{\mu A \ell}{6} \begin{bmatrix} 2 & 1 \\ 1 & 2 \end{bmatrix}, \quad \mathbf{K} = \frac{AE}{\ell} \begin{bmatrix} 1 & -1 \\ -1 & 1 \end{bmatrix}$$

where $\bullet$ is the density of the material, A is the cross section, ℓ is the length of the element and E the elasticity modulus. The equations for the two elements are assembled and after rearranging the terms to distinguish the actuated part from the unactuated, the complete equations have the following form:

$$
\begin{bmatrix}
\dfrac{\rho(A_1\ell_1 + A_2\ell_2)}{3} & \dfrac{\rho A_1\ell_1}{6} & \dfrac{\rho A_2\ell_2}{6} \\[2mm]
\dfrac{\rho A_1\ell_1}{6} & \dfrac{\rho A_1\ell_1}{3} & 0 \\[2mm]
\dfrac{\rho A_2\ell_2}{6} & 0 & \dfrac{\rho A_2\ell_2}{3}
\end{bmatrix}
\cdot
\begin{bmatrix} \ddot{q}_2 \\ \ddot{q}_1 \\ \ddot{q}_3 \end{bmatrix}
+
\begin{bmatrix}
\dfrac{\mu(A_1\ell_1 + A_2\ell_2)}{3} & \dfrac{\mu A_1\ell_1}{6} & \dfrac{\mu A_2\ell_2}{6} \\[2mm]
\dfrac{\mu A_1\ell_1}{6} & \dfrac{\mu A_1\ell_1}{3} & 0 \\[2mm]
\dfrac{\mu A_2\ell_2}{6} & 0 & \dfrac{\mu A_2\ell_2}{3}
\end{bmatrix}
\cdot
\begin{bmatrix} \dot{q}_2 \\ \dot{q}_1 \\ \dot{q}_3 \end{bmatrix}
$$

$$
+
\begin{bmatrix}
\dfrac{A_1 E}{\ell_1} + \dfrac{A_2 E}{\ell_2} & -\dfrac{A_1 E}{\ell_1} & -\dfrac{A_2 E}{\ell_2} \\[2mm]
-\dfrac{A_1 E}{\ell_1} & \dfrac{A_1 E}{\ell_1} & 0 \\[2mm]
-\dfrac{A_2 E}{\ell_2} & 0 & \dfrac{A_2 E}{\ell_2}
\end{bmatrix}
\cdot
\begin{bmatrix} q_2 \\ q_1 \\ q_3 \end{bmatrix}
=
\begin{bmatrix} 0 \\ F_1 \\ F_2 \end{bmatrix}
$$

Applying the linearizing feedback (3) the above equations become

$$\ddot{q}_1 = v_1$$

$$\ddot{q}_3 = v_2$$

$$\ddot{q}_2 = \mathbf{J}[v_1 \quad v_2]^T + R$$

where

$$\mathbf{J} = -\frac{1}{2(A_1\ell_1 + A_2\ell_2)}[A_1\ell_1 \quad A_2\ell_2]$$

$$R = -\frac{3}{\rho(A_1\ell_1 + A_2\ell_2)}\left[\frac{\mu}{3}(A_1\ell_1 + A_2\ell_2)\dot{q}_2 + \frac{\mu A_1\ell_1}{6}\dot{q}_1 + \frac{\mu A_2\ell_2}{6}\dot{q}_3\right.$$

$$\left. + E\left(\frac{A_1}{\ell_1} + \frac{A_2}{\ell_2}\right)q_2 - \frac{A_1 E}{\ell_1}q_1 - \frac{A_2 E}{\ell_2}q_3\right]$$

As it can be verified, the linear system is controllable which means that it can be driven with continuous linear feedback law. Indeed, condition $\mathbf{R}(\mathbf{q}^e, 0) = \mathbf{0}$, $\forall \mathbf{q}$ does not hold.

Set $x_1 = q_2$, $\mathbf{x}_2 = [q_1 \quad q_2]^T$, $x_3 = \dot{x}_1$, $\mathbf{x}_4 = \dot{\mathbf{x}}_2$

Using Proposition 1 we can prove that the system is second order non-holonomic. Notice how the columns in the sequence of matrices defined in Proposition 1 give exactly the vector fields which were to be calculated for the P. Hall basis.

Lets calculate first the vector fields that are formed with up to one level of bracketing:

$$\tau_0 = [\dot{q}_2 \quad \dot{q}_1 \quad \dot{q}_3 \quad R \quad 0 \quad 0 \quad 1]^T$$

$$\tau_1 = [0 \quad 0 \quad 0 \quad J_1 \quad 1 \quad 0 \quad 0]^T$$

$$\tau_2 = [0 \quad 0 \quad 0 \quad J_2 \quad 0 \quad 1 \quad 0]^T$$

and continuing with one level of bracketing:

$$[\tau_0, \tau_1] = [\frac{A_1 \ell_1}{2(A_1 \ell_1 + A_2 \ell_2)} \quad -1 \quad 0 \quad 0 \quad 0 \quad 0 \quad 0]^T$$

$$[\tau_0, \tau_2] = [\frac{A_2 \ell_2}{2(A_1 \ell_1 + A_2 \ell_2)} \quad 0 \quad -1 \quad 0 \quad 0 \quad 0 \quad 0]^T$$

Continuing to higher levels yields:

$$ad^2_{\tau_0} \tau_1 = [0 \quad 0 \quad 0 \quad 0 \quad \frac{3A_1 E(3A_1 \ell_1 \ell_2 + 2A_2 \ell_2^2 + A_2 \ell_1^2)}{2\rho(A_1 \ell_1 + A_2 \ell_2)^2 \ell_1 \ell_2} \quad 0 \quad 0]^T$$

$$ad^2_{\tau_0} \tau_2 = [0 \quad 0 \quad 0 \quad 0 \quad \frac{3A_2 E(3A_2 \ell_1 \ell_2 + 2A_1 \ell_1^2 + A_1 \ell_2^2)}{2\rho(A_1 \ell_1 + A_2 \ell_2)^2 \ell_1 \ell_2} \quad 0 \quad 0]^T$$

$$ad^3_{\tau_0} \tau_1 = \frac{3A_1 E(3A_1 \ell_1 \ell_2 + 2A_2 \ell_2^2 + A_2 \ell_1^2)}{\rho(A_1 \ell_1 + A_2 \ell_2)^2 \ell_1 \ell_2}[0 \quad -\frac{1}{2} \quad 0 \quad 0 \quad \frac{\mu}{2\rho} \quad 0 \quad 0]^T$$

$$ad^3_{\tau_0} \tau_1 = \frac{3A_2 E(3A_2 \ell_1 \ell_2 + 2A_1 \ell_1^2 + A_1 \ell_2^2)}{\rho(A_1 \ell_1 + A_2 \ell_2)^2 \ell_1 \ell_2}[0 \quad -\frac{1}{2} \quad 0 \quad 0 \quad \frac{\mu}{2\rho} \quad 0 \quad 0]^T$$

Since vector fields $\tau_0, \tau_j, [\tau_0, \tau_j], ad^2_{\tau_0} \tau_1, ad^3_{\tau_0} \tau_1$ are independent, the system is second order nonholonomic. Now, notice how these vector fields appear explicitly in the extended matrix of Proposition 1:

$$L_1 = \begin{bmatrix} \dfrac{\partial R}{\partial \dot{x}} \\ 0 \\ 0 \\ \dfrac{\partial R}{\partial x} + \dfrac{\partial R}{\partial \dot{x}_2} \cdot \dfrac{\partial R}{\partial \dot{x}} \\ 0 \\ 0 \\ 0 \end{bmatrix} \begin{bmatrix} J_1 & J_2 \\ 1 & 0 \\ 0 & 1 \end{bmatrix} =$$

$$= \begin{bmatrix} 0 & 0 \\ 0 & 0 \\ 0 & 0 \\ \dfrac{3A_1E(3A_1\ell_1\ell_2 + A_2\ell_1^2 + 2A_2\ell_2^2)}{2\rho\,\ell_1\ell_2(A_1\ell_1 + A_2\ell_2)^2} & \dfrac{3A_2E(3A_2\ell_1\ell_2 + A_1\ell_2^2 + 2A_1\ell_1^2)}{2\rho\,\ell_1\ell_2(A_1\ell_1 + A_2\ell_2)^2} \\ 0 & 0 \\ 0 & 0 \\ 0 & 0 \end{bmatrix}$$

$$L_2 = -\begin{bmatrix} A_2 \\ 0 \\ 0 \\ B_2 \\ 0 \\ 0 \\ 0 \end{bmatrix} \begin{bmatrix} J_1 & J_2 \\ 1 & 0 \\ 0 & 1 \end{bmatrix} =$$

$$= \begin{bmatrix} -\dfrac{3A_1E(3A_1\ell_1\ell_2 + A_2\ell_1^2 + 2A_2\ell_2^2)}{2\rho\,\ell_1\ell_2(A_1\ell_1 + A_2\ell_2)} & -\dfrac{3A_2E(3A_2\ell_1\ell_2 + A_1\ell_2^2 + 2A_1\ell_1^2)}{2\rho\,\ell_1\ell_2(A_1\ell_1 + A_2\ell_2)} \\ 0 & 0 \\ 0 & 0 \\ \dfrac{3\mu A_1E(3A_1\ell_1\ell_2 + A_2\ell_1^2 + 2A_2\ell_2^2)}{\rho^2\,\ell_1\ell_2(A_1\ell_1 + A_2\ell_2)^2} & \dfrac{3\mu A_2E(3A_2\ell_1\ell_2 + A_1\ell_2^2 + 2A_1\ell_1^2)}{\rho^2\,\ell_1\ell_2(A_1\ell_1 + A_2\ell_2)^2} \\ 0 & 0 \\ 0 & 0 \\ 0 & 0 \end{bmatrix}$$

In this case the partial derivatives of R are calculated only once, at the beginning of the procedure, whereas the direct calculation of the Lie brackets of the vector fields requires the calculation of the Jacobians of the vector fields involved.

6. Material Constraints

The degrees of freedom of the manipulated object may be subject to constraints. These constraints can arise from material strength limitations and/or obstacle avoidance requirements. While the latter apply directly on the object degrees of freedom, the former are usually expressed in the form

$$\acute{o} \le \bar{\acute{o}}$$

where $\acute{o}$ is the stress tensor of the structure and $\bar{\acute{o}}$ is the maximum admissible stress specified for the particular material and object. With some algebraic manipulation the above relation can be translated in terms of the finite element model node displacements. If one recalls the well known expressions that relate the deformation $\mathring{a}$, the displacement U, and the stress $\acute{o}$,

$$\mathring{a} = \mathbf{D} \cdot \mathbf{U}, \qquad \acute{o} = \mathbf{E} \cdot \mathring{a}$$

along with the element interpolation functions

$$\mathbf{U} = \mathbf{N} \cdot \mathbf{q}$$

the material constraints can be expressed in terms of the degrees of freedom. These constraints can be included in (10) with the use of Kuhn-Tucker multipliers:

$$\mathbf{M}\ddot{\mathbf{q}} + \mathbf{C}\dot{\mathbf{q}} + \mathbf{K}\mathbf{q} = \mathbf{B}\mathbf{u} + \left(\frac{\partial \mathbf{r}}{\partial \mathbf{q}}\right)^{T} \grave{\mathbf{i}}$$

$$\grave{\mathbf{i}}^{T}\mathbf{r}(\mathbf{q}) = 0, \qquad \grave{\mathbf{i}} \ge 0.$$

In this case, the multipliers cannot be eliminated, because they correspond to inequality conditions which do not reduce the dimension of the state space. The constraint terms in the above equations can be included, though, in the elastic and gravity forces terms as follows:

$$\mathbf{m}_{11}\ddot{\mathbf{q}}_1 + \mathbf{m}_{12}\ddot{\mathbf{q}}_2 + \mathbf{c}_1\dot{\mathbf{q}} + \mathbf{k}_1(\mathbf{q},\grave{\mathbf{i}}) = 0$$

$$\mathbf{m}_{21}\ddot{\mathbf{q}}_1 + \mathbf{m}_{22}\ddot{\mathbf{q}}_2 + \mathbf{c}_2\dot{\mathbf{q}} + \mathbf{k}_2(\mathbf{q},\grave{\mathbf{i}}) = \mathbf{b}(\mathbf{u})$$

$$\grave{\mathbf{i}}^{T}\mathbf{r}(\mathbf{q}) = 0, \qquad \grave{\mathbf{i}} \ge 0.$$

When the stress conditions are satisfied, $\grave{\mathbf{i}}$ vanishes and the equations describe the motion of the unconstraint system.

It should not be attempted however to determine the existence of nonholonomic constraints using the above equations. One the one hand, when constraints are respected, the equations coincide with (10). If the constraints are violated, it may be too late to examine controllability. On the other hand, material constraints do not correspond to the natural behavior of the system but are rather externally imposed conditions which are stated in order to confine deformations and avoid, perhaps, fracture. They are primarily used for optimization purposes. At any case, the results of the preceding sections may no longer hold if the inclusion of the material constraints destroys the linear structure of the elastic term.

7. Conclusion

A deformable object under manipulation can be modeled using finite element. This way, the distributed parameter model of the original system is converted to a finite dimensional underactuated mechanical system. In this light, the class of objects under study can be extended to include systems with combination of rigid and deformable objects, unactuated joints, flexible links and joints, rolling contacts etc.

This modeling method reveals a set of dynamic constraints. Depending on their type, several conclusions about the behavior of the model under control strategies can be drawn. For each type, the deformable object model is treated accordingly. Apart from these dynamic constraints, however, there could be additional ones that can relate to material strength limitations and/or obstacle avoidance requirements. A way to incorporate these constraints into the object model is described.

The study of deformable objects in the framework of underactuated mechanical systems indicated the existence of second order nonholonomic constraints. For the identification of this type of constraints there exist some heavy mathematical tools from nonlinear control. The model derived from the finite element methodology however, is linear and one should not be obliged to use such complex techniques. For this purpose, an alternative methodology has been developed, which does not require special mathematical skills and knowledge, and is also more computationally efficient than the original method. This method is illustrative with a simple example.

References

[1]   Tanner H G, Kyriakopoulos K J 1998 Modeling of multiple mobile manipulators handling a common deformable object. *Journal of Robotic Systems*, (15) 11: 599-623.

[2]   Sun D, Shi X, Liu Y 1996 Modeling and cooperation of multiple mobile manipulators handling a common deformable object. In: *Proc. of the 1996 IEEE Int. Conference on Robotics and Automation*, Minneapolis, Minnesota, pp. 2346-2351.

[3]   Terzopoulos D, Fleischer K, 1988 Deformable Models. *The Visual Computer* (4):306-331.

[4]   Kosuge K, Sakai M, Kanitani K, Yoshida H, Fukuda T 1995 Manipulation of a flexible object by dual manipulators. In: *Proc. of the 1995 IEEE Int. Conference on Robotics and Automation*, pp. 318-322.

[5]   Wu J, Luo Z, Yamakita M, Ito K 1996 Adaptive hybrid control of manipulators on uncertain flexible objects. *Advanced Robotics* (10) 5:469-485.

[6]   Yukawa T, Uchiyama M, Inooka H 1996 Stability of Control System in Handling of a Flexible Object by Rigid Arm Robots. In: *Proc. of the 1996 IEEE Int. Conference on Robotics and Automation*, Minneapolis MN, pp. 2332-2338.

[7] Tanner H G, Kyriakopoulos K J 1999 Analysis of Deformable Handling. In: *Proc. of the 1999 IEEE Int. Conference on Robotics and Automation*, Detroit, Michigan, pp 2674-2679.

[8] Spong, M W 1998 Underactuated mechanical systems. In: Siciliano B, Valavanis, K P, (eds) 1998 *Control Problems in Robotics and Automation, Lecture Notes in Control and Information Science 230*, Springer, pp. 135-150.

[9] Boothby M W 1986 *An Introduction to Differentiable Manifolds and Riemannian Geometry*, 2nd ed. , Academic Press Inc.

[10] Reyhanoglu M, van der Schaft A, McClamroch N H, Kolmanovsky I 1996 Nonlinear control of a class of underactuated systems. In: *Proceedings of the 35th IEEE Conference on Decision and Control*, Kobe, Japan, pp. 1682-1687.

[11] Campion G, D' Andrea-Novel B, Bastin G 1991 Modeling and state feedback control of nonholonomic mechanical systems. In: *Proceedings of the 1991 IEEE Conference on Decision and Control*, Brighton, England.

[12] Nijmeiher H, van der Schaft A J 1990 *Nonlinear Dynamical Control Systems*, Springer-Verlag.

[13] Rao S S 1989 *The finite element method in Engineering*, Pergamon.

[14] De Luca A, Mattone A, Oriolo G 1996 Dynamic mobility of redundant robots using end-effector commands. In: *Proc. of the 1996 Int. Conference on Robotics and Automation*, Minneapolis, MN, pp. 1760-1767.

Chapter 5

Applications and Industrial Experiences

Section 5.1

Simulation of Non-Rigid Materials Handling

C. Rizzi, M. Bordegoni, and G. Frugoli

Abstract. Many industrial processes are concerned with the manufacturing of non-rigid products or of automated systems that manipulate them. In this report, we present some testing cases and applications requiring the simulation of non-rigid material automatic handling. The testing cases have been carried out within the framework of international projects funded by the European Union. They are related to different industrial applications (e.g., automotive, aeronautics, and clothing) and require the simulation of different types of non-rigid materials, such as fabric, wire and foam. To this extent, we have used a software environment called SoftWorld, which allows the designer to model and simulate handling robots integrated with the static and dynamic behavior of the material. Besides, we have developed a non-rigid materials simulator integrated with haptic devices for modeling and simulating the touch of non-rigid objects.

1. Introduction

Today computer systems are widely used for designing and testing products. Current CAD and CAPE systems are proven to be effective in design the automated machinery and simulate the production process. They are especially useful in designing rigid products. However, several industrial sectors involve only the manufacturing of non-rigid products or the design of automated systems that manipulate them. In such a context, we have developed and integrated a software prototype for physically-based modeling and simulation, mainly serving for industrial applications rather than for computer animation [1]. The system, called *SoftWorld*, allows the user to interactively model non-rigid objects and to simulate their behavior [2]. SoftWorld adopts a physically-based model, known as particle-based model [3]. The model is force-based, and allows the description of a deformable object by means of a set of mechanical elements (particles), and a set of inter-particle forces that represent the macro-behavior of the material (e.g., elasticity, plasticity). Physical parameters characterizing the dynamic behavior are derived from experimental tests, e.g., for fabrics the Kawabata Evaluation System measures the physical data through bending and shearing tests [4].

The system is integrated with a kinematics work-cell simulator (Figure 1), in order to obtain an environment to design automated handling systems and simulate all the operations involving non-rigid materials, such as picking and transferring. While the kinematics simulator computes the trajectories of the robot parts, the non-rigid material simulator calculates the effects of the gripper fingers on the shape of the flexible object. Combined simulations of robotic system and the deformed material are the resulting simulation (Figure 1).

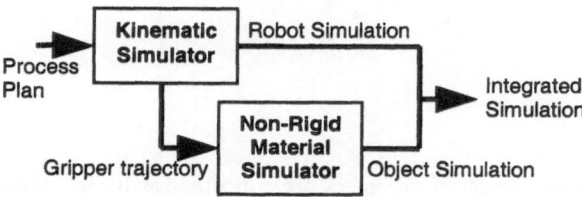

Figure 1. Architecture of the integration of the non-rigid material simulator with a kinematics robot simulator

Another issue we are working on is the effect to introduce haptic devices, which will give the users the possibility to touch the virtual products in order to perform some tests and analyses. For example, a designer can review a product design by evaluating the fitting between parts by checking the capability to assemble and disassemble some components, etc. It would also be possible to simulate a virtual mock-up so that human factors can be realistically simulated, and the users are able to check the usability of switches, knobs, etc. Another interesting application is using the simulation for training humans, such as physicians and surgeons, to perform tasks requiring sensori-motor skills. These are only some examples of the possible applications where touch can be introduced as complementing sensory channel to the commonly available vision channel. A requirement shared by several of these applications is the capability to interact with flexible objects through haptic devices.

In summary, the prototype integrated with the kinematics simulator and the haptic devices can be used as:

1. *Supporting the automation of industrial processes*: to model and simulate flexible industrial products (e.g., cables, fabrics, food, etc.), to study the final

configuration of the complete handling systems and to simulate automated handling processes;

2. *Aid the designer* during her/his task, for example in clothing industry, for the design of new base garments;

3. *Developing applications for training*, for example in maintenance area, through the integration of the system with haptic devices.

The experiments, presented in the following, are examples for the applications 1 and 3. Simulations rely on non-rigid material models that are experimentally validated and approximate the real object behavior with a good level of precision.

2. Applications

The testing cases described in this section have been carried out within the framework of international projects funded by the European Union, such as Brite/Euram projects NORMAH, SKILL-MART, MASCOT, and DMU-FS.

The non-rigid material simulation system has been tested on several applications involving the simulation of different types of deformable bodies. We started modeling flat objects, like fabrics for textile and clothing industry applications, and then we extended *SoftWorld* to handle solid objects for applications in different areas, as shown in Table 1.

	Automotive & Aerospace	Chemistry	Clothing	Food
Automation of industrial process	Simulate automated handling of wires/cables	Simulate automated handling of bag filled with liquid	Design and evaluate gripping systems Simulate handling operations of garment parts	Simulate automated handling of dough-like products
Support to designer's activity	Car soft-top design Simulate wires/cables under operating conditions		Virtual atelier: design new base garments Simulate garment drapability	

Table 1. Testing cases

The first application, we will describe, is the simulation of a typical operation performed in textile industry and to be automated: folding a cotton sleeve. This case study consisted of both studying gripping systems and of simulating the handling process to evaluate the gripping systems and the handling strategies.

Another set of testing cases is concerned with industrial operations requiring the manipulation of wires and cables. First experiment is about the automatic wire laying; the second one deals with the simulation of the action of a robot gripper, and the last one is the simulation of wire handling executed by two co-operating robots.

Another experiment concerns with a packaging application, where the handling devices must be tested to verify if an operation can be properly executed without slipping or damaging of the product. The system was applied in grasping and manipulating a bag filled with liquid, which is approximated as a model of foam material.

The last example deals with the use of the non-rigid material simulator integrated with haptic devices for modeling and simulating the touch of non-rigid objects. The user grasps a wire to put it into a new location. This allows us to check the model of the non-rigid object and the task to perform.

2.1 Fabric Manipulation

This application involves designing and testing an automated handling system for clothing manufacturing purpose, specifically, folding a lawn cotton sleeve. Two tests have been carried out: the design and simulation of specific grippers for fabrics, and the simulation of the handling system taking the behavior of the cotton sleeve into consideration.

The first test is actually a *virtual* test to find out different grasping configurations for two types of grippers (a *telescopic star gripping system* and an *X gripper*) specifically designed for fabrics [5]. We have tested, by means of an off-line programming system, the performance of both grippers in grasping generic shaped fabrics. At this stage, we do not consider the physical properties of the handled material: we still use the human skill to find out feasible grasping configurations [6]. Further considerations include simulating the entire handling operation and considering the physical properties of the non-rigid object.

Regarding the second test, the main objective is to qualitatively evaluate the approach and performances of the prototype. According to the model adopted in SoftWorld, the cotton sleeve is represented as a set of particles interconnected by internal forces describing the physical behavior (Figure 2).

The physical parameters used for material characteristics were derived from studies on non-rigid material categorization carried out by Brite/Euram Project NORMAH at University of Bristol. The environment consisting of a PUMA robot and a gripping system with two attachment points designed at Aristotelian University of Thessaloniki, has been modeled and simulated using the off-line programming system. We investigated, using simulation, different grasping and folding strategies in order to identify those causing the material breakage or the presence of creases during and after completion of considered operation.

Figure 3 shows steps for the simulated operation performed after having defined a proper handling strategy and the correct grasping points.

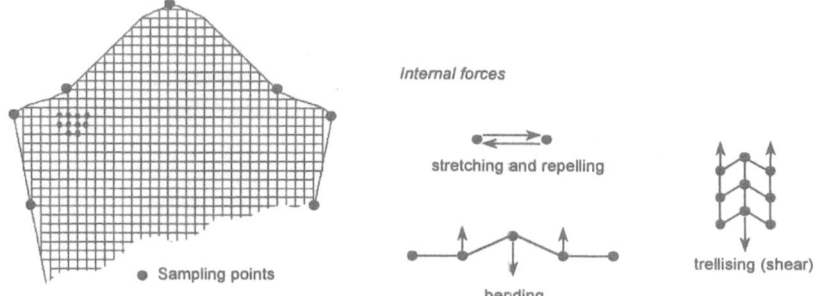

Figure 2. Particle-based model of the sleeve

Figure 3. Four steps of the cotton sleeve folding

Once the proper strategy is identified, we use the handling equipment from Aristotelian University of Thessaloniki to execute the same operation. Then we compare its result simulation by identifying some sampling points on the sleeve (Figure 2). We compared the positions assumed at specific time-steps of the simulation and during the real experiment. It can be seen that simulation results fit quite well with those of the physical experiment.

2.2 Wires and Cables Handling

The simulation of the behavior of manipulated wire can be carried out at different levels of accuracy (and computational cost) depending on the purpose of the simulation. The following experiments and different models representing wires are presented according to their specifications.

The first test is the simulation of laying wire, an operation that is becoming more and more important in the automotive and aerospace industry. In this example, we want to grasp a wire at one of its ends and move it along a given trajectory. The object is therefore constrained to follow the trajectory imposed by the gripper. Surrounding environment represents obstacles. Our focus is on the behavior of the wire during the operation.

The layout used for the test consists of:

- A wire laying on a surface;

- A gripper holding one end of the wire;

- Obstacles represented by thin walls.

The wire is modeled as a cylindrical surface: its discretisation is performed positioning the particles on the intersection of the isoparametric curves of the surface. Therefore, the model consists of a series of connected rings of particles distributed along the central axis of the object (Figure 4).

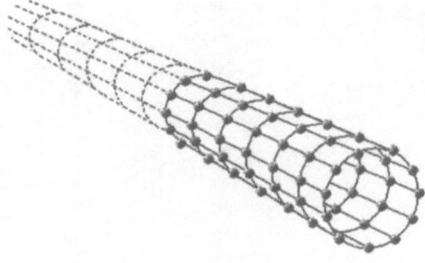

Figure 4. Surface-like model of a wire

Between every pair of rings, we set elastic forces to simulate the resistance to pulling action and compressive stress. Other forces are also set in order to simulate the bending resistance and the cross-section deformation at tension.

A trajectory constraint is imposed during simulation, based on the following assumptions:

- A grasped object, if the operation is correctly executed, can be considered as a single unit with the gripper;

- Material deformations generated by the gripper action can be ignored.

The trajectory constraint imposes a motion law to some particles. In order to define the constraint, we must determine the area involved in the grasping operation and the particles contained in this area. They are constrained to follow the gripper's trajectory. Therefore, the grasping operation is simulated imposing a predefined trajectory to the particles involved.

Figure 5 shows steps of the wire's behavior when manipulated as described. In the steps, we can see the behavior of the wire when colliding with obstacles.

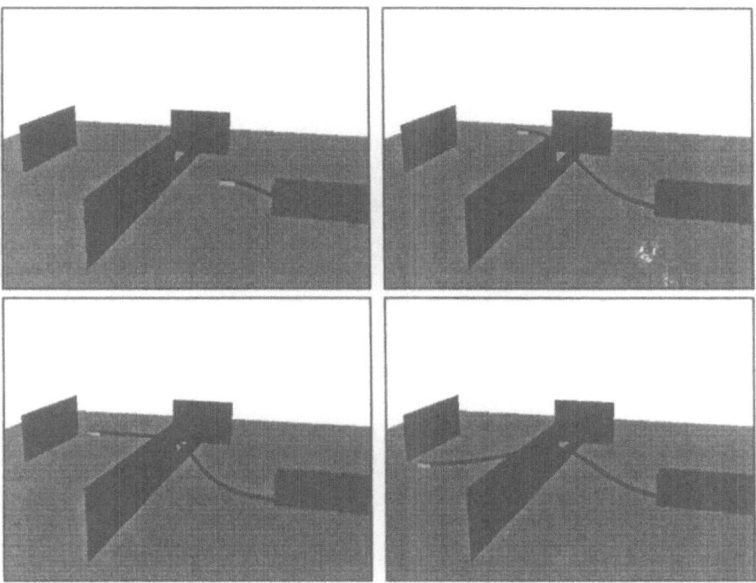

Figure 5. Wire laying simulation

The second experiment concerns with the simulation of a robot gripper grasping a wire. This example focuses on the deformation of the material due to the external forces applied by the gripper. A three-dimensional model of the wire is more suitable for this simulation since it offers a higher level of realism. The wire is modeled displacing particles radially within the cross-section as shown in Figure 6; by this way, a set of inter-particles forces is arranged in a symmetric way within the section.

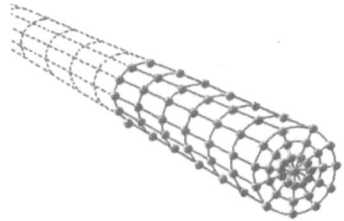

Figure 6. 3D representation of the wire: radial-based mesh

Two flat plates, whose surfaces are characterized by a roughness value, are used as the model of the gripper. The rigid objects defining the gripper must follow a given trajectory in the space, and when they are in contact with the wire, a mechanical compression and friction on the wire surface occurs. The gripper grasps a wire laying on a surface, and then the gripper can be seen translating the wire along Z-axis. The simulated action is simpler than the previous one, yet the models of the objects in this case are more complex. Figure 7 shows the wire behavior when manipulated as described: the deformation of the wire can be due to gripper's action.

Figure 7. Simulation of a wire grasped by a two-finger gripper

One of the main issues in automation in the aeronautical field is the automatic wire placing and cable bundling. The non-rigid material simulator together with the kinematics simulator are used to implementing a simulation of wire handling. The system includes a two-arm robot system equipped with end-effectors and three wires lying on a surface. The three wires are modeled using the surface-like model, as previously described, representing three different types of wires. The two-arm robot picks one of the wires up. The other two wires are then lifted up consequently. Figure 8 shows steps of the simulation. The simulation shows the behavior of the wires when handled according to a planned trajectory. This allows us to test and choose the appropriate wire handling task plan used in the actual robot system.

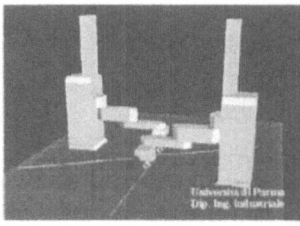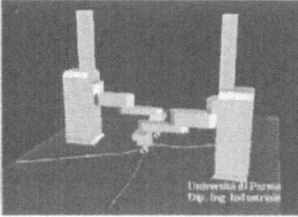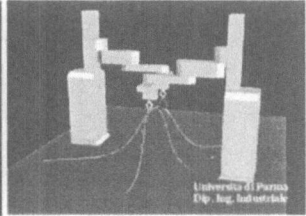

Figure 8. Sequence of simulation of a wire handled by a two-arm robotics system

2.3 Soft Bag Handling

Packaging tasks are common for many industrial applications, such as food industry, pharmaceutical sector, etc. One of the main goals for packaging non-rigid products is to execute the task without loosing or damaging the product. Simulation of the task allows us to predict the object behavior and possibly correct the handling task plan. The integrated system has been used for implementing the grasping simulation of a soft bag. The system consists of a two-finger gripper and a soft bag. The soft bag is used as an approximation of an object like a bag filled with liquid. The two-finger gripper grasps the soft bag and lifts it up. The simulation shows the behavior of the bag when handled according to a planned grasping trajectory. In this case, simulation allows us to test and choose the appropriate handling task plan to be used in the real robotics system. Figure 9 shows steps of simulation.

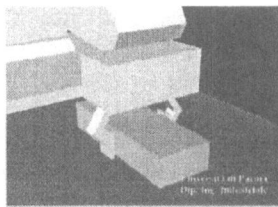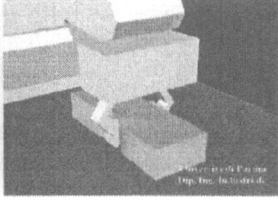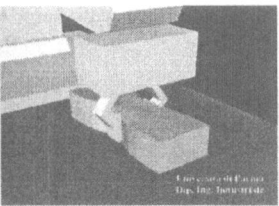

Figure 9. Sequence of simulation of a bag grasped and lifted up

In order to simulate the complete packaging task, we also need to simulate that the gripper grasps the bag and put it into a box. Figure 10 shows a sequence of images regarding to the simulation of this task. The non-rigid material simulator is also capable of computing the forces applied during the bag handling. These values are used by the end-effector in the real task execution.

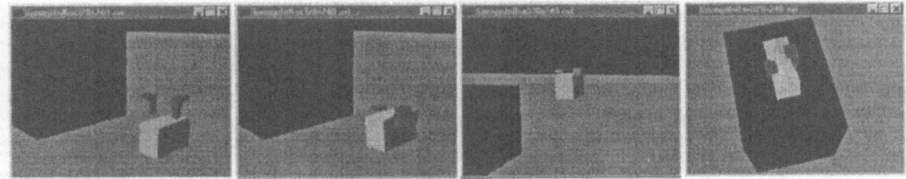

Figure 10. Sequence of pictures showing the bag packaging

3. Handling Operations with Haptic Device

Other special applications of object handling involve touching objects visually represented on the screen. Haptic devices that provide the users the sense of touch virtual objects offer this possibility. We have developed some applications of including haptic rendering of flexible objects, where the modeled soft objects can be not only seen but also touched [7]. The basic idea is to use the simulator to realistically compute both the forces generated by flexible objects against finger pressure and the deformed shape of the object, in order to perform an accurate haptic rendering together with a congruent visualization of the deformations. The test cases we have realized have been developed using two PHANToM haptic devices [8].

Some tasks such as placing an object in a given position, inserting a part into another one, etc. may require to be tested in advance. The visual simulation of the task is not always sufficient to test all the possible situations. An experiment we have developed is the haptic grasping of a flexible pipe. This is an application of placing the pipe in a certain position, and of extracting the pipe from a part where it is fixed in, etc. Figure 11 shows an example of manipulation of a pipe [9]. The two spheres are the visual representations of the users' fingers. The pipe shape is altered when it is touched.

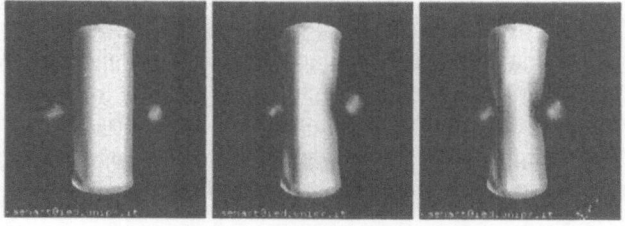

Figure 11. Steps of handling task of a pipe.
The user wears two haptic devices to feel the modification of the pipe

4. Conclusions

New software tools are essential for modeling non-rigid objects and for simulating their behavior during handling processes. In this report, we described some experiments related to the automatic handling of non-rigid materials that were carried out using a prototype developed at the University of Parma, Department of Industrial Engineering. It allows a system engineer to model and simulate the dynamic and static behavior of non-rigid objects manipulated by robot systems.

The tests described in the report allow us to verify the approach adopted (i.e., the particle-based model) and potential of the system. The testing results are encouraging and they demonstrate that the prototype and the approach are effective for representing and evaluating the behavior of non-rigid materials for a wide range of industrial applications, from automotive to food and clothing industry.

Another topic is the interaction with the non-rigid materials models via haptic devices; it is proven that haptic technology is very promising and useful for a variety of applications. The preliminary results of the experiments are considered positive and encouraging.

Acknowledgements

The authors would like to thank the EU Commission for funding our projects in this area, and our colleagues that participated to the research projects on non-rigid materials modeling and simulation.

5. References

1. Colombo G, Frugoli G, Rizzi C 1997 A computer aided system based on physically based modeling to study flexible products behavior. *In: ICPE 97*, pp 853-857

2. Denti P, Dragoni P, Frugoli G. Rizzi C 1996 SoftWorld: A System to Simulate Flexible Products Behaviour in Industrial Applications. In *European Simulation Simposyum (Ess 96), Vol 2*, pp 235-239

3. Witkin A 1995 Particle System Dynamics. In: *Siggraph Course Notes, Vol. 34*

4. Kawabata S 1980 The Standardization and analysis of hand evaluation. Textile *Machinery Soc. of Japan, Osaka*

5. Karakerezis A, Ippolito M, Doulgeri Z, Rizzi C, Petridis V 1994 A flexible gripping system for handling irregular flat non-rigid materials." In: *Euriscon 94 Conference, Vol. 3*, pp 1365-1378

6. Cugini U, Dragoni P, Ippolito M, Rizzi C 1998 Modeling and simulation of handling machinery integrated with dynamic and static behavior of non-rigid materials. *IEEE Robotics and Automation Magazine, Special Number on "Robotics and Automation in Europe: Projects funded by the Commission of the European Union*, 48-56

210

7. De Angelis F, Bordegoni M, Frugoli G, Rizzi C 1997 Realistic Haptic Rendering of Flexible Objects. In: *The Second PHANToM User Group Workshop,* Cambridge

8. SensAble Technologies, Inc. http://www.sensable.com/

9. Rogalla O, Dillmann R, Frugoli G, Bordegoni M 1999 A General Approach for Simulating Robots for Flexible Material Handling. In: *ISATP 99,* pp 211-218

Section 5.2

Robotics for Deheading White Fish

R. Buckingham and A. Graham

Abstract. This chapter provides an overview of work conducted over almost a decade into the automated handling and processing of white fish with the specific objective of increasing fillet yield whilst deheading.

1. Introduction

Robotic technology for the handling of food products is generally absent from factory floors. This is primarily due to the peculiarities that handling food adds to the general pick and place task for which most robots are designed. These include the complexity of handling non-rigid products that are both easily damaged and infinitely variable in shape; the hygiene requirement which stipulates IP65 or better for the hose down environment; and the reality that the food industry produces low margin products that only make significant profits at large volumes.

In this environment, tasks that have both positive commercial indicators and are technically achievable are rare. The fish processing industry is no exception.

This chapter considers a single example called "Robofish" in which a team of designers developed a totally new piece of equipment for the fish processing industry.

The chapter describes the technical achievements and the process by which the project reached its current pre-production stage. It is now almost 10 years since the project was first discussed as a funding proposal for submission to the European Commission. The project subsequently received

3 years funding as a research and development project under ESPRIT (Robofish I) and a further 2 years funding as a Demonstrator Action again under ESPRIT (Robofish II). It is clear that without such funding this project would never have been conceived.

Robofish I

The objective of Robofish I was to introduce off-the-shelf robotics into a fish processing line. However it became clear that the combination of weight, reach and speed meant that existing industrial robots would not be suitable. The first significant change was to embark on the development of a new twin armed continuously rotating robot. This robot was built and was demonstrated collecting fish from a conveyor and delivering the fish to the deheader. The cost of the robot was kept down by recognising that the robot did not require micron accuracy or repeatability. This is true of the vast majority of food applications.

Having built the robot, and having impressed a number of user companies, it then became clear that such a system would be unlikely to work on a production floor. This resulted in a complete review of the technical specification in order to meet the project's technical and operational objectives.

Robofish II

The second phase of Robofish was intended to produce a ruggedised Robofish I. However the makeup of the project team, which included 8 user companies and only 2 technology developing companies, ensured a different future. Issues such as price, reliability, maintenance and continuous quality of the final result came to the fore.

This review resulted in a comprehensive re-design of the system. Where Robofish I used a vision system, Robofish II used mechanical guiding and proximity sensors. Where Robofish I moved each fish 2m from conveyor to blades and then released the fish just when control is vital, Robofish II was designed to minimise the movement of the fish whilst controlling the position of the head and the body with respect to the knives through the cutting process. In short Robofish 1 created an industrial type robot, whereas Robofish II created a task specific manipulator.

The final result, as described below, does not look like a robot. In fact Robofish became a self-feeding deheading machine. However, this should not belittle the technology within. The mechanisms that enable the fish head and body to be controlled throughout the cut whilst achieving a through put of 1 fish a second are extremely innovative. Part of the innovation is to incorporate features that enable the machine to withstand the harsh environment - the chemical washdown sprays and expected heavy handed operation.

Deheading Fish

The initial concept was to utilise robotics in order to improve the deheading yields in fish processing plants. This was the one consistent factor during the development of Robofish; namely the belief that a machine could be developed that would produce a minimum yield increase of 1%.

Deheading is typically the first process that is conducted on land. A 1% loss in the first process towards producing the high value added final product equates to between 2 and 5% loss downstream. The 1% improvement target is probably also an under-estimate of what can be achieved since even now it is still not possible to provide a reliable measure of current deheading yields. This is due to the different measures used across the industry and the variations in the product due to season, sex, fishing grounds etc.

However, it was agreed at the outset that a 1% increase in yield over existing machines would provide a viable payback, without taking into account the more intangible benefits that such a system would bring.

The processing environment for Robofish is a typical land based fish factory within a fishing area. The fish are caught and gutted onboard the trawler and loosely packed in ice before being unloaded at the dockside factory. The boxes of fish are emptied onto conveyors that bring the fish to the first operation, that of deheading the fish. The purpose of this task is to separate the high value fillet from the low value head. The perfect outcome of the process is to leave the maximum amount of meat on the fillet, whilst ensuring that no part of the fish fins or head skeleton is included within the fillet. The latter would need to rectified by time consuming rework of the fish.

Existing machines that achieve the best fillet yield require careful manual positioning of the fish by the operator. However the working conditions are such that an operator cannot grasp, transfer and accurately place a fish weighing between 1 and 12 kg in these machines for long periods. Hence the machines in current use have been designed to simplify the operators' task in order to reduce fatigue and injuries, but at the cost of reduced yields.

2. Robofish I – Initial Research and Testing

The objective of the first project funded by the EC was to interface a robot with a high yield deheader and to optimise the delivery of the fish to maximise yield. The robot was to copy the function of the operator.

The manual task can be simplified as follows:
1. Locate, grasp and remove the fish from the conveyor.
2. In parallel with #1, assess the fish for damage and illness.
3. Translate and rotate the fish so that the fish is fed into the deheading machine with the head held at the correct height and angle to the vertical so that the deheading machine knives cut at the base of the fish head. Once the fish is held within the deheading machine release the fish, before loosing fingers. A typical deheader is shown in Figure 1.
4. Return to collect next fish.

Figure 1   Baader 421 Deheader

The key requirements of the automated process were therefore as follows. At a rate of one fish every two seconds:

1. Locate the fish using computer vision techniques (in the visible spectrum).
2. Grasp the fish in a manner that does not damage the fillet.
3. Transfer and place the fish to within ±2mm in the deheader.
4. Release the fish.

3. Cod

The fish under consideration are fresh cod of mass between 1kg and 12kg and of length up to 1200mm, Figure 2.

Figure 2   Standard fish

Cod belong to the white fish family and are the most difficult fish to handle. Red fish are more uniform in shape and tend to lie more regularly, and flat fish are simpler again. The cross section of cod can be described as being a blend of triangular and oval, although, just like humans, fish come in all sorts of shapes and sizes. Data collection showed that there are no close correlations between fish weight, length, distance from jaw to skull, eye to snout, or indeed any measurement that is guaranteed to enable a fish to be accurately modelled based on a small number of data readings. This, coupled with the natural oils present in the skin, the water present in all parts of the process and the neck cut introduced during gutting, means that if a fish is thrown onto a flat surface the fish can lie in almost any position, although there is a tendency for the fish to lie on one side. This raises the first technical issue of locating a fish with the accuracy required to enable the fish to be grasped reliably and released in the correct position. The tolerance required during deheading is ±2mm, so the issue of establishing a relatively fixed datum is vital. The flesh of the fish is unsuitable to establish any datums and although direct visualisation of the fish skeleton using x-rays was discounted there is a close correlation between certain surface features that can be identified from a visible spectrum image and the underlying skeleton.

The human approach of grasping the fish in a random grasp and then placing the fish accurately within the deheader by means of real time vision and force feedback was not considered possible to replicate, bearing in mind the huge product variations and the operating environment.

The slippery surface to the fish also removes non-adaptive gripping solutions that rely on friction. The ability of a robot to catch and manipulate a fish by means of surface contact only would be a major research project in its own right. Within Robofish I only gripping techniques that used the underlying skeletal features of the fish were shown to be effective. The fact that the high value fillet must not be damaged focused attention on the head. In fact, the head is ideal for grasping and delivery to the deheader, since the remainder of the fish skeleton has very few features that are easily identifiable or appropriate for grasping. This is fortunate since the head is the low value part of the product, and hence dramatic grasping techniques that lock onto the skeleton, even if they cause damage to the surrounding tissue, are not inappropriate.

Diagrams of the head skeleton, Figure 3, clearly show the features that are useful for locating and grasping the fish. These are covered with a variable amount of flesh, but the positions of the features are relatively easy to identify from surface views. In particular the eye sockets, the snout and the jaw bone are prominent.

Snout

Eye socket

Jaw

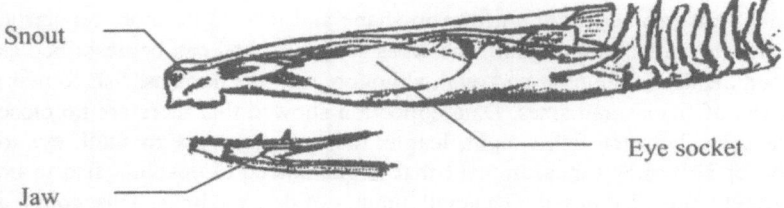

Figure 3   Fish head skeleton

Since cod have a tendency to lie on their sides the view perpendicular to the fish $Y_f Z_f$ plane was identified in the early stages as being suitable for a visual location paradigm. This view also encompasses the prominent skeletal features. The coordinate scheme attached to each fish is shown in Figure 4.

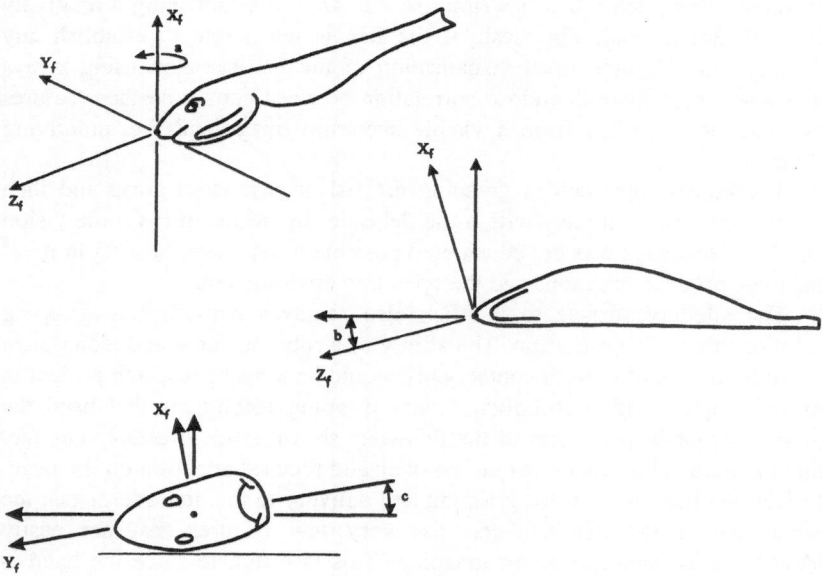

Figure 4   Fish coordinate scheme

The fish snout can also be used as a reference point and orientation of the snout can be approximated by fixing a second point along the skull, recognising the cod have relatively flat snouts. X_w, Y_w, Z_w and rotZ_w offsets can be defined relative to the fish head centre or fish snout in order to fix a grasp position.

4. "Whirling Dervish" Robot Solution

The task of collecting and feeding a fish to the deheader using a robot can be broken down into the following subtasks:

1) Identification of a fish and determination of its position and orientation relative to the conveyor at a point in time before collection. From this data a collection point can be identified.
2) Analysis of fish data to define a collection point.
3) Planning of the robot trajectory to collect and deliver the fish.
4) Movement of the gripper to track the fish a short distance on the conveyor before grasping.
5) Grasp the fish.
6) Transfer the fish to the deheader, by firstly sliding the fish off the conveyor and then carrying the fish into the deheader. The dynamic behaviour of the fish during this phase exerts large forces on the fish. The accelerations involved in rotating and transferring the fish exert considerable loads on the gripper.
7) Release the fish once the velocity of the fish is synchronised with the deheader.
8) Return the robot to the loiter position and wait for the next fish.

There are three technical elements to this solution that were developed within the project. These are the vision system to locate the fish, a gripper to grasp the fish and the robot to move the gripper between conveyor and deheader.

Identification of Fish Position

Many hundreds of measurements would be required to define the exact geometry of an individual fish on the conveyor. However the head can be reasonably defined by the six variables listed below. These dimensions are given in the fish coordinate reference frame, see Figure 4.

The Z_f and Y_f dimensions define the outer profile when viewed from perpendicularly above the conveyor.

The X_f dimension defines the thickness of the fish.

The angle 'a' defines the orientation of the fish about the X_f axis.

The angle 'b' defines the orientation of the fish about the Y_f axis.

The angle 'c' defines the orientation of the fish about the Z_f axis.

These measurements could be recorded by means of two vision system, one (called 'Vision 1') situated above the conveyor with a view perpendicular to the conveyor and the other with a view orthogonal to the direction of motion of the conveyor and the view of Vision 1. It was decided for simplicity to only implement Vision 1. This meant that only Z_f, Y_f, and [a] could be measured. The variation [b] was deemed small enough to ignore and X_f was found to be approximately proportional to the overall length of the fish. The angle [c] was an uncontrolled function of the delivery of the fish to the system and became a source of concern.

More specifically accurate information was available for: the X_w, Y_w position of the snout; the X_w, Y_w position of the centre of the visible eye socket; the X_w, Y_w position of the apex of the neck cut; and the X_w, Y_w position of the tail of the fish. Hence it is possible to define: the length of the fish; the X_w, Y_w head centre; the X_w, Y_w position of the snout; and the orientation of the fish about the Z_w axis. By analysis of the data it is also possible to estimate the likely thickness of the fish at the head centre, and the direction of rotation about the X axis (the sign of [c]). This process is critical to the success of grasping a fish.

The critical element of the image processing was the extraction of features that are required to control grasping. These are the lie of the fish and the position and orientation of the snout, Figure 5.

A Marel vision system was adapted and developed in order to capture the fish images. On a sample of 200 fish the results produced 100% success rate in finding the head of the fish, 95% success in identifying the snout of the fish and 99% success in fish lie detection.

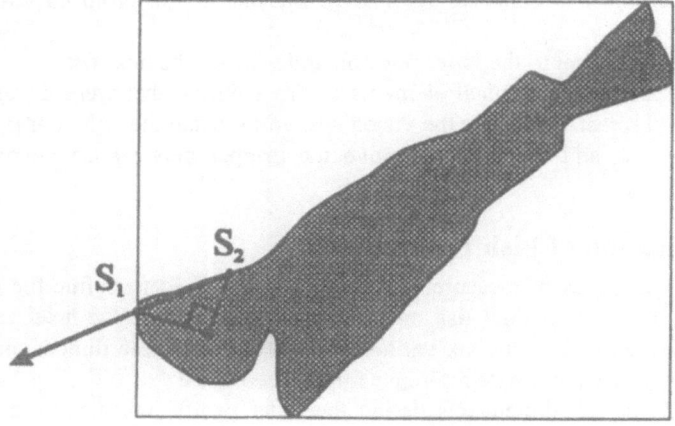

Figure 5   Binary fish image

Gripper

The basic function of the gripper was to establish a strong grasp on a fish which is arriving either head or tail first and either on its left or right side and then to be able to release the fish reliably at the correct point in the cycle.

The action of gripping the fish had to be fast because the cycle time was short. The action must also control the motion of the fish once some contact is made, before the secure grasp is achieved. When grasping a slippery fish, one result can be that the gripper closes but the fish ends up... somewhere else! Because the position of the fish must be known in order to correctly deliver the fish to the deheader the distance of the snout from the centre of grasp, when the grasp is secure, must be measured. Once the fish is securely grasped the fish is slid off the conveyor and transferred to the deheader. The closing force of the gripper was specified as 600N in order to keep the largest fish securely grasped whilst undergoing a 10g (!) acceleration. The dynamics involved in this system layout where a major cause of concern.

The release of the fish needed to be both fast and smooth so that the direction and orientation of the fish did not change. This implied that the gripper action had to release the gripping fingers so that the fingers did not snag on part of the skeletal structure of the fish.

The gripper implemented was the patented two fingered snout gripper which comprised one fixed finger, called the skull plate and a two jointed finger called the jaw wedge. The jaw wedge had one controlled joint and one compliant joint. The controlled motion was that of rotating the jaw wedge towards the skull plate. The principle of this grip is very similar to the human hand with the thumb opposing the first three fingers. When this human grip is applied to the head of a fish the natural configuration is for the three fingers to form a slightly concave plate, with the thumb rotating and locking into the jaw folds of the fish. With the gripper in the fully closed position the jaw wedge pressed against the skull plate. A pneumatic cylinder acting along a line coincident with the gripper's major axis, provided actuation to the free finger through a crank and sliding joint mechanism.

The swivelling jaw wedge pierced and penetrated into the jaw folds at some point behind the jaw bone. The shape of the wedge caused a self-rotation about the end of the free finger. This rotation caused the fish to slide in the gripper until the rear edge of the jaw bone made contact with the wedge tip. At this point little further slip occurred and the wedge tip dug deeper into the head. The grasp was a combination of location behind the jaw bone, and a pressure grip between the upper and lower sides of the skull plate. The skull plate also provided some centralising effect.

This two finger design was necessary in order to grasp fish lying on either the left or right side without losing the specific design features of the skull plate and jaw wedge. The necessity to turn the gripper over for different fish meant that the gripper needed to be thinner than the thinnest fish. The smallest fish thickness with the 1-6kg range was measured as 45mm.

Since the fish can move during the grasping process the accuracy of the vision data is lost. A sensor was therefore incorporated within the gripper to measure the final position of the fish snout with respect to a known datum. This information was then used to control the exact height at which the fish enters the deheader. Non contact sensors were excluded for reasons of fish detritus build up, leaving an air-sprung linear potentiometer as the most robust solution.

The prototype gripper, shown in Figure 6, achieved a grasp and release rate of over 99%.

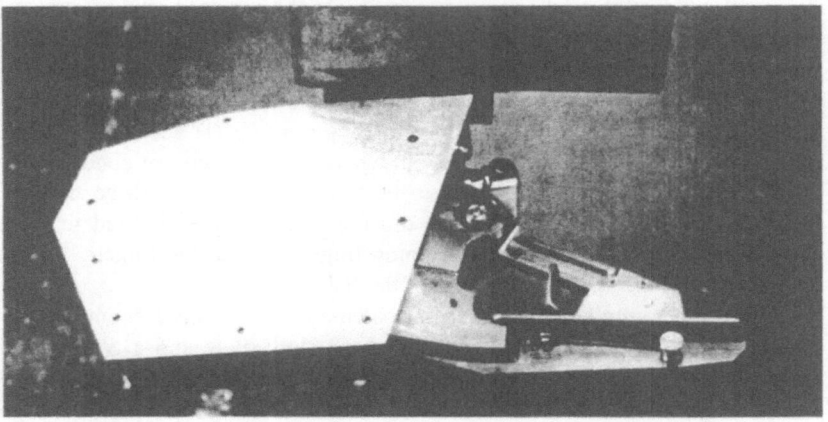

Figure 6   Prototype gripper

Robot

The task of the robot was to move the gripper through a cycle that was defined uniquely for each fish. The gripper trajectory was calculated from the data the vision system provides and the known location of the deheader.

An accuracy of ±2mm at the critical parts, modest by assembly-robot standards, was sufficient to optimise yields. But at the same time the robot had to be extremely fast with a cycle time of 2 seconds over a 2m diameter working area, with a payload including gripper of 8kg. It was latter discovered that fish weighing up to 12kg would need to be processed. The collection zone of the infeed conveyor and the size of the deheader determined the working area. In addition the motions required to pick up left- and right-lying fish, and to present them to the deheading blades with the centre line of the skull tilted back 60° relative to the backbone demanded high dexterity - in particular a 3 degree of freedom wrist. It was observed that there would be a useful saving in energy if the robot were capable of continuous rotation because, as in most food industry transfers, the velocity vectors of the arriving conveyor and the outfeed to the deheader are unchanging so the rapid reversal of the robot's motion is unnecessary. The collection and outfeed zones occupy a segment of an annulus subtended by an angle of over 180 degrees and it was easier to return to the loiter position before collecting another fish by continuing to rotate in the same direction.

No existing robot was found to meet this specification, let alone one constructed to food-grade standards. It was decided to design a new robot, in which the cost of obtaining high speed and a relatively large working volume would at least be partly offset by the reduced accuracy (about 20 times lower) compared with precision industrial assembly. It was decided that the robot should be a relatively conventional SCARA design not unlike a larger version of the Adept PackOne, but with all exposed surfaces in stainless steel or polymers, much reduced accuracy, and the possibility of continuous rotation at the shoulder joint.

The design was made such that two identical arms and grippers could be mounted on a single vertical column, with the two arms operating approximately 180° out of phase in order to double the fish collection rate, (from 4 seconds to 2 seconds cycle time). Both arms were capable of independent control by the vision system subject only to the need to avoid inter-arm collisions. The resulting design became known as the "Whirling Dervish". The schematic for the two arm design is shown in Figure 7.

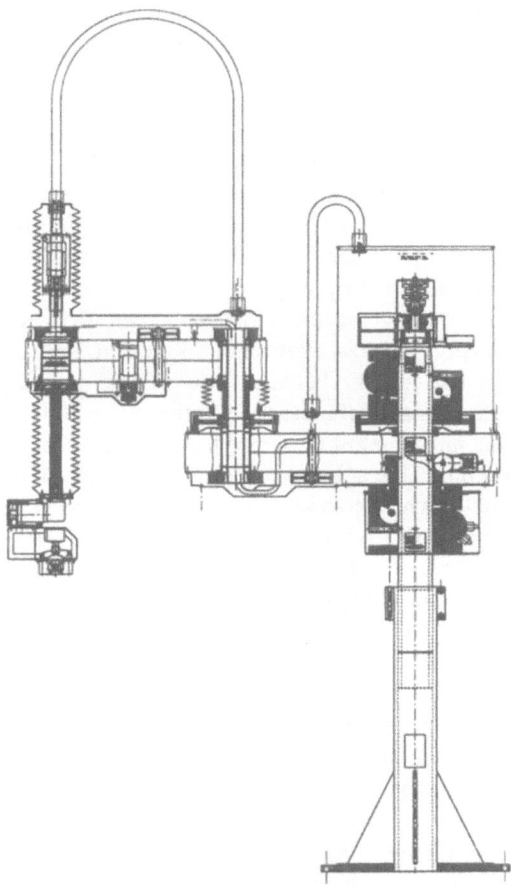

Figure 7 Robot schematic

In order to reduce the inertia of the arm the main motors of the arm, for the shoulder, elbow and yaw joints, were all mounted on the fixed central column of the design. Drive to the elbow axis was achieved via steel belts. This added to the complexity of the joint to encoder transforms but made the acceleration and accuracy targets achievable.

In order to mount the main motor gearbox combinations on the central column special annular spiroidal gearboxes were designed. The design aimed to maximise the inner diameter of the annulus in order to maximise the rigidity of the supporting column. An important feature of the spiroid design is the ability to control backlash to an acceptable level.

The arms themselves were a stainless steel monocoque (stressed skin) design, again to reduce weight and to provide IP65 surfaces.

The vertical axis of the arm was implemented with a proprietary two-function THK ballscrew-spline unit that incorporated the vertical motion and the yaw motion of the wrist.

Another major design feature was the rotating slip rings that enabled services to reach the distal joints of the arms. This rotating unit is the section at the top of the central column in Figure 7. In order to reduce the number of communication channels and power cables the transputer based joint controllers for the distal axes with their power amplifiers were mounted within the rotating system, requiring just two sliprings, one for communications and one for power.

The motion required at the wrist was well defined by the process. Collecting fish from the conveyor required the full 360° of yaw motion and the flipping of the gripper between one of two positions 180° apart to account for right side and left side lies. At the deheader the gripper needed to be flipped to the mid-point of these positions. Hence this axis only required a tri-state operation. This 'flip' axis was designed about a rotary vane pneumatic actuator which was secured in one of the three position by means of a further pneumatically activated locking piston. The third axis of the wrist was needed to offer the head of the fish to the deheader at the correct orientation to maximise the yield of the fillet. This axis required a maximum motion of only 90 degrees, and typically uses only 30 degrees of this motion.

The robot was fully sealed to achieve IP65 by means of the stainless steel monocoque and other rigid and flexible covers linked by flexible polymer bellows and sealed slip rings.

Part of the design specification was to achieve an overall accuracy of ±2mm. It was fortunate that the critical phases of the cycle all occurred at low velocities. Collection of the fish occurs at 500mm/s and the deheader infeed conveyor operates at only 150mm/s. This reduced the need for dynamic accuracy, although ±5mm was achieved. An absolute accuracy of within ±2mm was achieved with a repeatability of ±0.5mm.

Safety consideration for such a large and powerful robot were paramount in the design. No joint motion may commence unless all inputs to a safety ring interlock circuit are satisfied. These inputs monitor not only the interlock switches, relevant power supply rails etc. but also the electrical, communications and control software integrity. If the interlock is broken while movement is occurring, the main axis motors are all braked by disconnection from the drivers, short circuiting and by actuation of mechanical brakes under spring pressure. The combined braking torques were large enough to slow the end-effector from nearly 5m/s to standstill within 50mm or so, providing high intrinsic safety in possible collision involving other equipment or operating personnel.

Robofish I Results

The key results relate to the reliability of the system components and the ability to collect random fish from the conveyor and deliver them accurately to the deheading machine.

A successful grasp therefore included the correct operation of the vision system, the robot and the gripper. Similarly a successful delivery included the operation of the snout sensor, the gripper and the robot.

The vision system achieved greater than 98% successful identification of the fish lie and snout position. The same success rate was achieved by the robot and gripper combination. The results of the one and only low volume trial were encouraging but certainly not conclusive. From a series of 44 fish, 43 fish were successfully cut. 75% of these were in the typical ratio of head to total body weight, indicating that the cut is near optimal.

Figure 8   High quality 'V' cut showing improved yield

Figure 8 illustrates the qualitative nature of the yield gain. The fillet shows the 'V' cut which includes meat from the back of the head, whilst also showing no sign of skull bone being included. Other fish were cut such that bone was included, which would have required rework under production conditions.

The major conclusion from this first phase was that a robot could collect random fish from a conveyor and transfer the fish into a deheading machine. This in itself amazed the gathered user group and acted as a window onto future possibilities.

As Robofish II then demonstrated these achievements were shown not to be particular relevant in terms of technological solution, but they were vital in demonstrating that a robot could be associated with fish. The key outcome was that some progress had been made and that the fish processors could see in part what could be achieved.

5. Robofish II – Revised Objectives

The key objective of Robofish II was to allow a significant group of European fish processing companies, called the User Team, hands-on access to the Robofish development, to ensure that the technological advances could be brought to market as a viable commercial product by the Technology Developing partners - Marel (Iceland) and Oxford Intelligent Machines (UK). This was the specific objective of the funding received from the EC under the Demonstration Action heading. As partners the following fish processing companies were potential purchasers of the Robofish technology: Grandi, Iceland: Pieters Visbedrijf, Belgium: Royal Greenland, Denmark: Bluecrest Food Ltd, UK: Nestlé R&D, Sweden: Fiskiðjan Skagfirðingur, Iceland: Melbu Fiskindustri, Norway.

The specific objectives of this Demonstration Action were defined as:

- To produce a formal Customer Functional Specification for the production machine embodying all relevant requirements agreed by the User Team.
- To verify that it is feasible to attain yield improvement of 1 - 3% on large numbers of fresh fish when removing the head whilst still held in the machine's gripper.
- To improve and simplify the Demonstrator System to ensure that functionality, safety, reliability, and cost all conform insofar as possible to the Users' requirements.

The commercial objective remained the same as Robofish I - increase fillet yield in excess of 1%.

The change in management between Robofish I and II was significant. Firstly the number of technology partners was greatly reduced – only OxIM and Marel remained from the original group. Secondly, the large group of potential users was brought into the project to formalise a "Customer Functional Specification" and to assess whether the development would be commercially viable. This group was not interested in the detail of the design and was not involved in the day to day decisions. They did have a strong steering role and their views were actively sought especially after demonstrations.

The technology was to be based on the results of the original Robofish project. However, early feedback from the User Team reinforced a number of concerns regarding the viability of the particular robotic solution developed in Robofish I. This resulted in a change of focus from "vision guided robotic handling of fish" to the implementation of a "self-feeding robotic deheader".

The result was that none of the original hardware from Robofish was re-used and the process was completely redefined.

One of the main restrictions of Robofish I was that the robot fed an existing deheader which was designed for manual feeding. In particular the fish had to be released before the cut was made and as a direct result of releasing the fish the positional control afforded by the robot was lost.

The first months of Robofish II concentrated on the position and force control requirements for optimal yield - just the problems that Robofish I could not consider.

The complexity of the robot was directly related to unnecessary restriction imposed by the infeed and outfeed systems.

Working with one of the most experienced deheader designers, the team developed a new concept for deheading with the aim of maximising yield.

Video analysis of the manual action of feeding this deheader showed that the position and force control of the head of the fish is critical to achieving the optimum yield, and led to firm decisions as to optimal entry, intermediate and exit points for the cutting blades, relative to typical fish features.

This improved deheader clarified the aims of the project. It was clear that the small size of the new deheader made an inline feeding system much easier to implement. An in-line compact system led away from the SCARA robot design used in Robofish I towards an in-line robot motion in which the motion of the fish is reduced to an absolute minimum. Moving the fish through small distances meant shorter cycle times and reduced dynamic effects.

Another feature exposed by the tests was the need to control the head and the body of the fish during and after the cut. During the cut it is useful to have a static fish in which the weight of the body is not necessarily taken entirely by the fish's spine - since this can be severed during gutting. If the body is held after cutting it is also possible to consider automated infeed to a filleting machine – a crucial commercial issue. The optimum was to control the relative and absolute positions of the head and body separately. This changed the role of the gripper from being just an attachment method for the head to being a mechanism that controls the both the head and body of the fish.

6. Self-feeding Robotic Deheader

The machine built within Robofish II was a self-contained unit with all electronics and computing within IP65 cabinets. Figure 9 has been annotated in order to describe the process as follows:

1. Fish are placed individually on the infeed conveyor (1). Fish are placed head first lying on their right side.
2. As the fish rise up the inclined conveyor (2) the weight of each fish is measured.
3. This weight is passed to the machine controller (3). In parallel with the motion of the fish up the conveyor, all the necessary decisions are made regarding whether the fish is processed and if the fish is to be processed the unique cut path based on the fish weight is calculated.
4. Once the fish drops into the holding basket (4) all the decision making is complete. The holding basket conveyors are then accelerated forward (5), delivering the fish into the robot gripper, or in reverse (6) ejecting the fish from the machine and into a different process.
5. A sensor recognises the presence of the fish and the gripper (7) is closed.
6. The holding basket then opens (8) at the same time as the head control mechanism (9) and body control mechanisms (10) start to descend.
7. Once the cut has been made the fish head is released from the gripper (11) and the body is released by the body basket (12). The machine is designed to run at 20 fish per minute, although 30 fish per minute is achievable.

Figure 9   Complete machine (back side)

The machine was designed as a series of discrete sub-systems. This was in order to identify clear separation lines between the OxIM and Marel designed components and to ensure the functionality of the machine. In a production machine there is scope for much closer integration between certain sub-systems.

The major subsystems are described below.

Framework

The framework formed the mechanical interface between Marel/OxIM components. It was a simple, rugged, stainless steel structure - giving plenty of room for modifications. The framework defined the overall dimensions of the machine, excluding the infeed conveyor. The dimensions were approximately 3m in length, 1.2m in height and 1.5m in width at the widest point. The framework was clad with removable panels or hinged doors. The doors provided access to the machine for cleaning, whilst the panels could also be removed for maintenance purposes. Both panels and doors also fulfil a safety function. In a production machine the hinged door panels would be linked into the maintenance safety circuit - under normal operation opening a door would cause an emergency stop, but during maintenance these switches would be disabled.

The framework had two distinct sections. As shown in Figure 9, the left-hand side of the framework contained the body basket mechanism and the right-hand side of the framework contained the head gripper and actuator mechanisms. These sections were separated by a central bulkhead, on which were mounted the knives and clapper bars.

The head and body control sub-systems were mechanically discrete, and mounted to the framework using a small number of fasteners. The simple mechanical interfaces produced using this arrangement allowed for development of the sub-systems without any requirement for gross changes to other sub-systems or the framework.

Infeed Conveyor

The infeed unit, Figure 10, demonstrated one concept that could feed multiple Robofish units. The infeed unit was the only part of the prototype machine that was entirely separate. It did not locate with the framework and only had a control signal connection to the machine controller. It is likely that a simpler, more compact system would be used for a production machine. Infeed and outfeed requirements are typically unique for each application. One new feature was the incorporation of a catch-weigh station as part of the slotted, inclined tilted conveyor.

The role of the infeed was to signal to the controller that a fish is travelling and then to provide the weight information. The speed of the infeed could be altered through the controller.

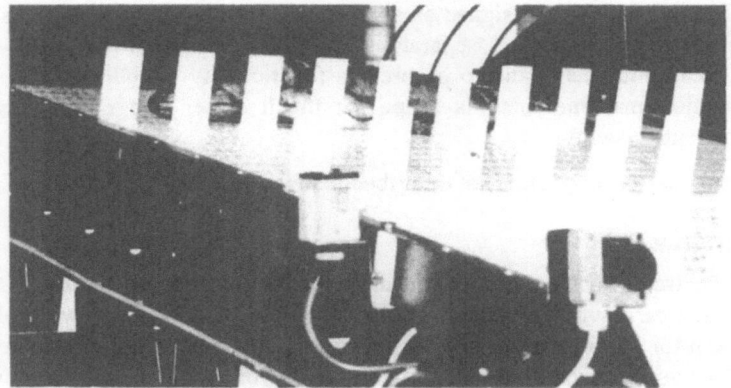

Figure 10   Fish on infeed

It is typical for the operator to be raised slightly as in Figure 11. Not shown in this diagram is the supply of fish. These would be provided from a standard conveyor system. The action of the operator would be to grasp the fish at the head and tail, rotate and translate by sliding to position the fish head first and if necessary rotate the fish about its longitudinal axis so that it is lying on its right side. By acknowledging that an operator would be almost essential to the correct operation of the system the decision was taken for the operator to ensure correct feeding of one (or two machines). This resolved (by avoidance) some of the more difficult problems that were identified in Robofish I – namely the delivery of fish to the system in a controlled and regular configuration. Robots are not good at dealing with large scale variations. Where possible it is much better to remove and/or simplify any sensor systems by constraining the variations. In the case of Robofish II the use of the operator in reducing infeed variation was critical.

Holding Basket

The holding basket acted as a buffer between the input conveyor and the machine. The holding basket is the mechanism on the top of the machine frame, Figure 11.

The holding basket had two actions: it located the fish and was capable of moving it forward to deliver the fish into the gripper; and in reverse to eject the fish from the machine.

As the fish dropped into the holding basket it was passively aligned and positioned so that the robot gripper could grasp the fish and take the fish through the cutting process. Based on fish weight, the decision whether to reject or process the fish was then received from the controller. For instance, the deheading process could be programmed to only process fish between selected weight values, those outside this range being re-routed to another processing line.

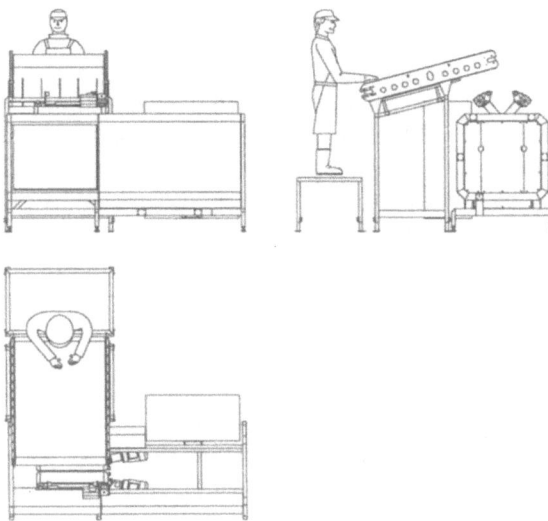

Figure 11   Infeed showing operator position

Clapper Bars and Knives

The clapper bars provided the means of locating the fish axially with respect to the blade planes during the cut. The blade drive arrangement was improved over the prototype systems by use of direct (straight-line) drives. The blades are industry standard blades, chosen to be capable of deheading the largest fish which will be handled.

Robotic Body Basket

The body basket mechanism carried the fish body during the cutting process. The fish entered the machine at the top on the machine centreline, and the separate body and head exited the machine at the bottom. Each body basket was a V shape, with the V angle chosen following the initial trials for the best support of the sides of the fish body.

The innovative design simplified the replacement of parts. Entire motor and gearbox assemblies were removable and could be replaced using a minimum of tools. Each axle assembly including the bearings and pulleys were designed to removable without tools for off-line maintenance or replacement.

Gripper

The gripper had perhaps the strongest genetic links with the Robofish 1 variant, but with a number of significant changes. An extra degree of freedom was added. Both degrees of freedom were pneumatically actuated. One cylinder caused the skull spikes (small horizontal spikes mounted at the ends of the gripper fingers) to be pressed into the side of the skull. The second cylinder moved the skull spikes down, to squeeze the head of the fish firmly into the V-wedge at the bottom of the mechanism. A rigid and reliable grip was achieved in this manner. The skull plate acted as the datum for fish head motion. Robofish I had demonstrated that a three point spiked gripper could maintain a firm grasp, but the requirement to flip the gripper and for the gripper to be thinner than the thinnest fish ensured this could not be pursued. Robofish II returned to this option since these restrictions were, by design, removed.

Head Actuator Mechanism

For the head actuator, a combination of 2 two degree of freedom linkages was used to produce a full three degree of freedom planar mechanism.

The gripper was mounted at the blade end of this mechanism and could therefore be moved in the required three planar degrees of freedom:

- Vertical - to match the vertical motion of the body baskets and to pass the fish through the static blades.
- Radial - in/out to push the fish against the clapper bars, and
- Rotational - to orientate the fish head at the correct angle with respect to the blades.

The radial (push) axis was responsible for applying a constant force along the axis of the gripped fish head and aligned the fish longitudinally.

One significant advantage of the mechanism eventually chosen was that the drive motor and gearbox were located away from the knives. This made observation of the cut during processing much easier and made cleaning and servicing quick and straightforward.

The robotic elements were controlled using a standard industrial servo controller. In this case a Baldor Optimised 8 axis board was used, with a combination of control code written in C running on a separate PC and Baldor's own language MINT running on the servo-control board.

The code required to calculate paths for the mechanisms was not trivial, although due to the planar decoupled design the kinematics, the code was compact and uncomplicated. Similarly tuning the machine to achieve maximum performance was straightforward – certainly less complicated than Robofish I.

The key aspect of the software was to optimise each cut based on the length and weight measurement of each fish. This objective resulted from the video analysis of manual techniques. Each fish is unique and therefore by treating each fish as a unique product gives the opportunity to maximise the yield from each fish.

User Interface

A major reason for the success of Robofish II was the parallel development of a new Marel controller. Robofish II was used as one of the first test cases for the M3000 controller. The Robofish II executable application running on the M3000 was responsible for the high level functionality and user interface of the machine. Robofish II was the first object-oriented C++ application that has been written for the new and powerful Marel M3000 controller.

The M3000 is a computer with an LCD display and a small key pad, housed within an plexiglass and stainless steel IP67 enclosure. The M3000 is the small silver box with the blue screen in Figure 12. The processor is a Power PC chip running the PSOS real time operating system. The M3000 has no hard disk since this function is fulfilled by Flash RAM and the networking capability. The M3000 is equipped with a large range of interface capabilities from serial and parallel ports, CAN and ethernet. With the power and interface capabilities the M3000 can run large real time processes.

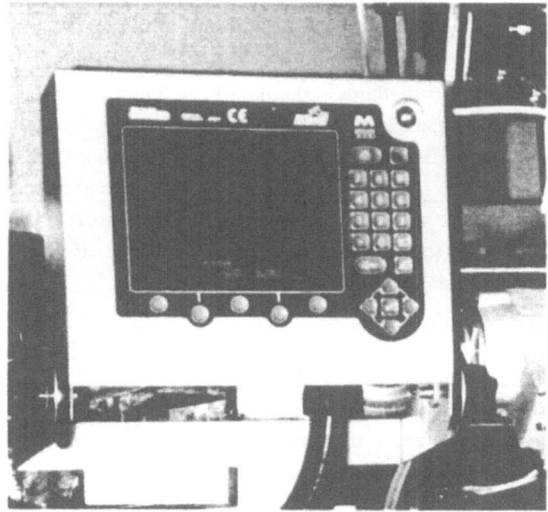

Figure 12   M3000

To the user the M3000 appears as a number of screens. Each screen ascribes certain functions to a series of function buttons and allows certain actions or access to other screens. The design of these screens is central to the operation of the User Interface.

7. Results and User Feedback

The key result from the Robofish II Demonstration Action was the demonstration of a new and novel piece of machinery for the fish processing industry, supported by results of machine trials that clearly indicate the commercial viability of the new machine.

The achievements of the machine are indicated below, and linked with the specification defined by the User Team.

Customer Functional Specification	Achievements
The machine should handle gutted, head on, cod, haddock, salmon and saithe.	The machine has been operated with cod, haddock and saithe.
The machine should offer an average yield improvement of at least 1% over the deheading machines commonly used today. Current yields are: cod 27-35%; saithe 19-24%; haddock 20-24%; salmon 9-11%	Saithe 19.8% (1st Stage Yield Trials). Cod 28.6% (2nd Stage Yield Trials).
The weight range should be 500g to 12.0kg, gutted with head, probably in two ranges – 500g to 2.5kg and 2.5kg to 12.0kg.	The higher weight range has been demonstrated.
The machine should operate at a minimum rate of 30 fish per minute.	20 fish per minute has been demonstrated.
Different deheading methods should be selectable for individual fish, to allow collar bones to be left on the fish head or body.	Yield results are for "collar bone with head deheading". Leaving the collar bones with the head has been demonstrated.
The machine should require one low skill operator, whose task should be to present the machine with fish that are roughly uniformly aligned.	Demonstrated. The fish are loaded all head-first on their right side.
The quality of the head cut should not depend on the operator's technical skills or processing experience.	Demonstrated
The machine must not cause damage to more than 1% of fish. Dropped fish should be recycled.	Not yet known.
The fish fillets must not be damaged causing a reduction in value.	Demonstrated.
The system should be for both land based and ship based operation. Size and integration with other machines have increased importance for ship based operation.	This is feasible, although the development to date has been for land based operation. This is the typical development route for new equipment, since onboard operation is particularly testing.
The machine must be assumed to be operated 90% of a shift in a cold, damp and wet environment, either on land or on ship.	Not yet known.
Its mechanical design should make the machine easy to clean and disinfect with the methods and chemicals that are currently used in the fish processing industry.	Demonstrated.

The machine should be fitted with the guards and covers that are necessary to prevent injuries to the operator, not only during normal operation but also during cleaning or maintenance operations.	Demonstrated.
A safety mechanism should prevent operation unless all the required covers and guards are securely in place.	The machine will be CE marked.
The machine will be hosed down every shift and cleaned every day.	Such cleaning has been conducted without problems.
The machine footprint should be approx. 2m by 2m.	Machine footprint is 3m by 1.5m. A production machine will be smaller.
The machine should be linked to a data acquisition system.	M3000 controller has CAN and ethernet links (as well as serial and parallel I/O)
The user interface must be simple but also informative.	Demonstrated.

Table 1   Achievements

2nd Stage Yield Trials Results

The yield trials were conducted with the Machine at Marel, using cod rather than saithe as used in the 1st Stage Yield Trials.

100 fish were processed in groups of 5 - i.e. fish were placed in five adjacent flights of the infeed conveyor, rather than waiting for each fish to be processed before loading the next. The weight range of 2.5kg to 7kg corresponded exactly with the range used to design the machine.

The yield results were as follows:

Weight	Length	Condition	Head (1)	Head (2)	Yield (1)	Yield (2)	Flesh Left
[g]	[cm]	[g]/[cm³] *1000	[g]	[g]	[%]	[%]	[%]
4031	76.9	88.1	1137	1102	28.4	27.6	0.8

Table 2   Yield Trials Results

Information supplied by the Users (in the Customer Functional Specification) indicated that typical yield for cod are in the range of 27% to 35%. The reasons for this wide range include season, gender, condition, and the variation caused by manual involvement.

These results indicated that the yields were at the low end of the range - 28.4%. In fact, the key figure is the 0.8% that remained with the heads. This was flesh that was removed using a sharp knife post-deheading and hence defines the optimum deheading yield for this batch of fish.

An average yield within 0.8% of the optimum manual deheading yield is a very strong indicator of an excellent yield performance. The cut quality was also of a consistent high quality.

8. Conclusions

Both Robofish I and II were highly successful in their own way. Robofish I was the essential pre-requisite for Robofish II. It was presumed that it was possible to use a robot to pick up fish. Robofish I demonstrated just that and enlightened a large number of users as well as most of the design engineers.

Robofish II was a highly successful project in a number of different ways.

First, the structure of the team that included a large number of Users and small focused group of Technology Developers was extremely productive. The input received from the User Team had a significant impact on the direction taken in the project. At the same time, having only two Technology Developers, Marel and OxIM, who had worked together previously on Robofish I, led to efficient management and clear objectives.

Having a diverse group of Users did provide a diverse range of opinion, so it was essential that the technical partners took final decisions. In fact, the Users preferred to work as consulting experts to the project uninvolved in the day-to-day management of the project.

Second, a prototype self-feeding robotic deheading machine was built that has a clear market niche and proven commercial advantages. Many of the machine features are novel and a comprehensive patent has been filed. The impact of such a machine, in which IT plays a central role, will be a profound stimulus to the industry. There are very few machines in the Fish Processing Industry which process each product as a unique individual and can provide comprehensive information on the each product and the performance of the process in real time. The Robofish machine is designed to be part of a continuous process with a batch size of one.

Whilst being recognised as a prototype, the machine demonstrates many of the features that are needed for operation under factory conditions.

Third, the issue of yield improvements has been central to project. It is clear from the results of the machine that the target of a 1% improvement over current automated deheading systems has been demonstrated. This was endorsed by the User Team. This yield improvement is significant because it provides a quantifiable commercial payback within an environment where to maximising a limited global natural resource has commercial and political implications.

These achievements are considerable. Europe is already highly competitive in the global market of fish processing equipment, but the industry has only recently started to take onboard the opportunities made available through IT. Although the machine built under Robofish II is only a single machine, deheading is a process that is critical for every factory, world-wide. The machine is conspicuously different from all other machines and offers dramatic yield improvements to the extent that the machine could change the direction of fish processing technology development. As such there is a significant opportunity for future growth and development of the European fish processing equipment industry.

The impact of the machine will also be felt in other sectors of the food processing industry. There are very few examples of automated machines working on factory floors that process non-rigid, slippery, highly variable product such as fish.

It is clear that the User Team considers that these are valid points. Of course, the true test of this is whether the machine is taken up by the market. The return on investment is well quantified based on a cautious sales prediction. However the general assumption is that the new self-feeding robotic deheader developed under Robofish II is a potential market leader.

Acknowledgements

The authors wish to acknowledge the following specific people who were involved in Robofish I :

Eric Andersson (Matcon); Ingvar Kristinsson (IceTec); Koorosh Khodabandehloo, David Fisher, Neil Maddock, Andy Shacklock, Lesley Barry and Geoff Newell (Univeristy of Bristol); James Foley, Peter Davey, Tim Jones, Mark Barlow and Ian Treherne (OxIM Ltd); Hordur Arnarson, Jon Benediktsson and Petur Snaeland (Marel hf); Gudmunder Einar Jonsson (Grandi hf); Karl-Heinz Robrock and Frank Cunningham (EC) and the Review Team: John Gray (University of Salford), Franky Demeester (European Independents) and Anita Geogeghan (Microchem Ltd) .

And in Robofish II:

Hordur Arnarson, Petur Snaeland Jon Benediktsson, Siggy (Marel hf), Peter Davey (OxIM) the User Team, and Karl Heinz Robrock, Tor-Ivar Eikaas (EC) and the Review Team, John Gray (University of Salford), Anita Geoghegan (Microchem Ltd) and Francky Demeester (European Independents).

About the Authors

Dr Rob Buckingham rob@ocrobotics.co.uk was the Project Manager for the University of Bristol during Robofish I and then during Robofish II worked under subcontract to Marel hf after the formation of Oliver Crispin Consulting Ltd to act as the Project Secretary and software developer.

Mr Andrew Graham was chief designer on Robofish II whilst working for Oxford Intelligent Machines Ltd.

Section 5.3

Application of LLW Robots to Distribution Lines

Y. Maruyama

Abstract. Kyushu Electric Power Co., Inc. began developing robot technology in 1984 as a method of performing outage-free live-line work (LLW). The outage-free technique allows power companies to maintain a stable, uninterrupted power supply to customers, even during distribution line maintenance such as pole replacement. This also allows for tasks to be completed in a more comfortable, safer working environment. In 1989, the first stage robot, Phase I, was developed and was then followed by the completion of Phase II in 1997. These robots are intermediate steps toward our future goal: development of a fully automatic LLW robot. It has been shown that the development of Phase I and Phase II improve working conditions in difficult and dangerous live-line operations. The robots pioneered the development of mobile outdoor robots in Japan and around the world, and are anticipated to be widely used in the electric sector, the railway industry, and other sectors worldwide. This report outlines the history of our LLW robot development, its new developments, and its state-of-the-art field applications.

1. Background of LLW Robot Development

1.1 Outage-free Maintenance Technique

Outage, including pole replacement, wire installation, and pole mounted transformer installation has been unavoidable for conventional distribution line work in Japan. Until recently, the direct LLW technique had only been applied to partial, simple maintenance. In direct LLW maintenance, linemen wear rubber gloves and handle live-lines directly. As overhead lines are often located near pedestrian areas or trees, accidents such as electric shocks and grounding faults inevitably occur. Since the early 1970's, the use of insulated, covered distribution lines has been promoted nationwide. But maintenance of these lines still requires linemen to remove line covers manually, again resulting in accidents such as touching live-lines. In addition, linemen, who wear heavy insulated garments for protection, have faced terribly uncomfortable working conditions during Japan's hot and humid summer. The direct LLW maintenance has these serious problems, making a new method much required. (Figure 1.1)

In the 1980's, dependence on electricity continued to increase along with industrial advancement and social progress. This trend has placed further demand on outage-free maintenance carried out by power companies, also increasing the need for LLW. With this increase in demand, Kyushu Electric shifted to full use of the indirect LLW technique in 1988 to improve workers' safety. This technique, in which linemen perform line maintenance using insulated hot-sticks, without touching live-lines directly, brought a drastic decrease of electric shocks.

Figure 1.1   Manual LLW maintenance

1.2 Concept of LLW Robot Development

Although the use of hot-sticks has improved workers' safety, handling the sticks for a long time is strenuous and several workers are required. In order to improve work efficiency, we decided to introduce robot technology into live-line maintenance. Robot technology has been developed in deliberate, progressive steps since 1984 as an outage-free LLW technique. The development of a fully automatic LLW robot is challenging, with many issues to be tackled. Thus, we decided that robot technology would be developed in the following three phases. (Figure 1.2)

- **Phase I** - Bucket operation type LLW robot (manual mode)
  An operator stands in a bucket mounted on the boom of a truck, and controls the manipulators involved in the robot manually using visual information.

- **Phase II** - Ground operation type LLW robot (semi-automatic mode)
  A "semi-automatic" robot is defined as one in which tasks are shared between a human operator and robot. The robot manipulators are remotely controlled by an operator who sits in a cabin inside the truck on the ground. Featuring several automatic functions developed in Phase I, this robot has achieved significant labor savings and consistent, high quality work performance. Simple operations such as identification of work objects and their locations, the determination of operation starting time and its ending time, and the evaluation of performed tasks are commanded by the operator. Phase II, which is a transient model to fully automatic maintenance, has shown more satisfactory work efficiency than expected. This efficiency has been realized by combining human skills and the accuracy of the robot.

- **Phase III** - Ground operation type LLW robot (fully automatic mode)
  This robot will execute LLW in a fully automatic fashion by identifying work objects and tasks, and by evaluating work results afterwards.

	Phase I (present model)	Phase II (currently being put into practical use)	Phase III (future)
Working conditions	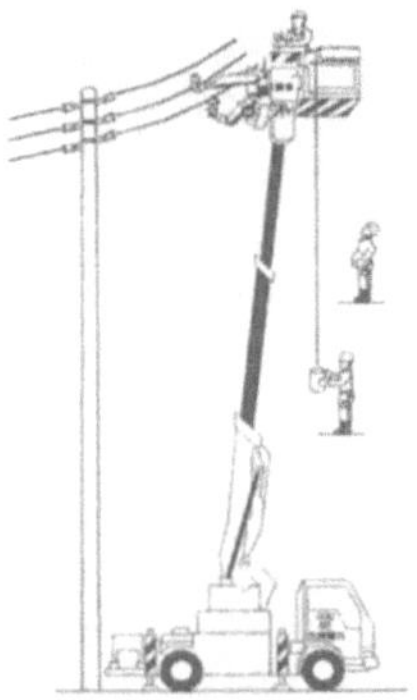		
Safety	Danger of falling from poles: low	No danger	No danger
Work environment	Work skills: intermediate	Work skills: low	No work skills required (fully automatic)
Workers required	3 persons	2 persons	1 person

Figure 1.2   Steps in robot development

2. Development of Phase I

2.1 Phase I Prototype

In March, 1986, the prototype version 1 of Phase I, a single manipulator robot, was created after a competition between two wire companies. (Figure 2.1) Serious problems were found after the evaluation of the prototype. The following improvements were then made:

- Replacing the single manipulator with dual arm manipulators. The single manipulator robot had to juggle work objects back and forth between pole and wire, which caused unstableness during windy conditions;
- Reducing the truck size to fit the narrow roads in Japan;
- Partially automating work for higher efficiency;
- Analysis of 90 maintenance operations has proved that all operations can be performed by 10-15 multi-purpose tools.

The final version of Phase I was created after incorporating the above improvements. (Figure 2.2)

Figure 2.1   Prototype of Phase I

Figure 2.2   Final version of Phase I

2.2 Application to Field Work

Phase I was completed with the development of prototype version 6 in 1989 through joint development carried out by three manufacturers: Yaskawa Electric Corporation, and two wire companies. Phase I robots were distributed to all 85 customer service offices within three years after completion, and the application to field work started after the exercise of robot operations.

The sophisticated movements of the robot manipulators have drawn an enthusiastic response from operators and workers. The use of robots in field work has increased to 90% or higher for maintenance work throughout the company.

3. Development of Phase II

3.1 Major Features

The development of Phase II, a follow-up to Phase I, started in 1990. This robot, an outdoor mobile unit, is operated semi-automatically and features all-weather electrical insulation for the first time in the developmental process. Its major functions are described below. (Figure 3.1)

3.1.1 The Third Arm

While Phase I required temporary mobile support during operation, Phase II was designed to complete LLW tasks without requiring support, to reduce costs and to improve maneuverability on narrow roads in Japan. At this point, the third arm, which works both as a temporary wire support and as a device for suspending heavy objects, was mounted on Phase II. Developments focused on designing a system compact and light enough to mount onto a single truck.

Three years later, all necessary functions were successfully mounted onto prototype version 3, along with the third arm. This arm was designed to suspend heavy objects including materials and equipment, and to support live-lines temporarily during maintenance. A powerful compact hydraulic system was employed for the third arm as high-speed motions were not required for maintenance.

242

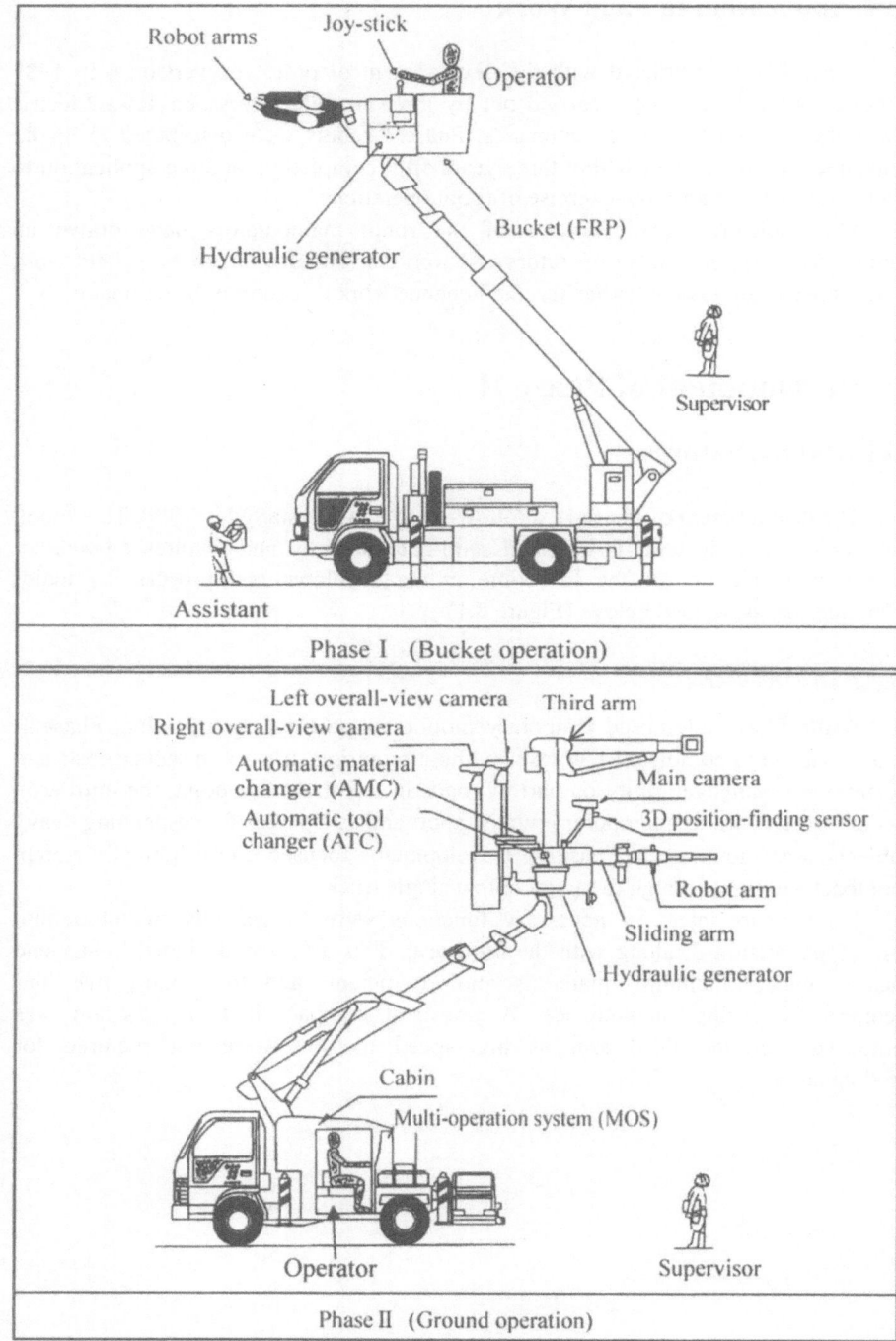

Figure 3.1 Overview of Phase I and Phase II

3.1.2 Sliding Robot Arm

Most distribution lines in Japan are installed above the ground at the roadside. Since wires are commonly laid horizontally in the present pole framing structures, robot arms often cannot reach the wires from the roadside. Guiding a robot from the opposite side of the road reduces the efficiency of the maintenance. In some cases, the guidance itself is impossible due to physical interaction between the robot and the environment. A back-and-forth sliding motion for the bucket is thus integrated on the Phase II. This function enables all wire work to be performed from the roadside. With this modification, all line tasks can now be handled efficiently from a single roadside working position.

3.1.3 Seven-Axis Arm for Obstacle Avoidance

In LLW tasks, robot arms should perform wire operations without physical interference between the arms and the environment, such as poles and wires. This implies that the arms should have seven axes to avoid obstacles during the wire operations. We thus have introduced robot arms with seven axes and have put the arms into practice for the first time.

3.1.4 End-Effectors for Remote Control Operation

Tools suitable for each task were necessary to automate the overall work process. A complete analysis of all tasks in distribution line work has shown that the following tools are needed for the maintenance. (Figure 3.2)

- 8 single-function tools including a hose-free compressor (15t)
  This compressor does not require high-pressure hydraulic hoses which are not appropriate for automatic operations.

- 3 multi-function tools including a wire peeler/polisher
  This tool enables workers to peel and to polish insulated wires at the same time.

- 3 supporting tools including a protective pipe feeder*
  This tool enables continuous feeding of protective pipes automatically.

- Automatic tool changer (ATC) and automatic material changer (AMC)
  Since the operator controls robot arms at a remote site, tool exchange and material delivery should be performed automatically. Tools and materials are set in predetermined locations. According to operation programs, tool exchange and material delivery are achieved at the end points of the robot arms. Long materials, such as protective pipes, are suspended by the third arm.

---

*protective pipe: insulated pipe for electric wires used near buildings being constructed to prevent electric shocks or grounding accidents

Wire end stripper   Wire cutter

Compression tool   Wire binder

Wrench (right angle)   Wire stripper

Figure 3.2   End-Effectors for Phase II

3.1.5 Robot Vision

Automatic robot operation accomplishes the following:
(1) recognition of work object;
(2) measurement of the location of objects in 3D space;
(3) movement of the robot arm to the starting point of operations;
(4) automatic operation in the correct order.
Procedure (1) is still undergoing research around the world. Kyushu Electric has

introduced cameras so that a human operator can recognize work objects at a remote site. Procedures (2), (3) and (4) were automated more easily. A 3D position-finding sensor using laser measurement technology was mounted on Phase II. This sensor has enabled the robot to recognize the spatial relationship between the robot and work objects. First, a laser is projected onto work objects, and images of the objects are measured by cameras. From the difference between a laser-projected image and a natural image, work objects are identified and the distance between the robot and the objects can be computed. (Figure 3.3)

An operator monitors a screen display via cameras, traces the motion of the robot, and sends work commands to the robot or executes auxiliary manual operations if necessary. Introduction of artificial eyes into distribution work had been regarded as impractical technology before the development of Phase II. Kyushu Electric has achieved automatic operations with high accuracy ($\pm$2.0mm/1 - 5m) by introducing the artificial eyes. (Figure 3.4)

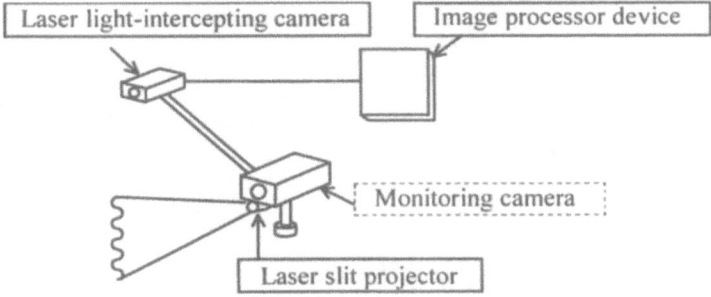

Figure 3.3   3D position-finding sensor

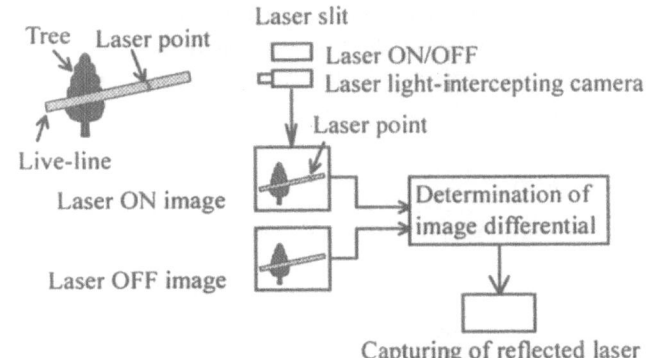

Figure 3.4   Distance measurement using laser

3.1.6 KYUDEN Language

Robot operation requires exclusive software programs for individual tasks. In the initial stage of Phase II development, Yaskawa and the two wire companies developed software programs with different programming languages. In practical maintenance, individual tasks are related to one another. This implies that the developments of individual task programs are strongly related and using different languages will hinder further developments.

In order to improve the efficiency in developing the software, Kyushu Electric has designed a robot language specialized for distribution line work. This language has the capability of specifying tools, materials, and work procedures in a block-diagram manner using task-specific terms. The design of the language is based on the work analysis of field engineers. The design was completed in 1994 and has been applied to prototype version 3. (Figure 3.5)

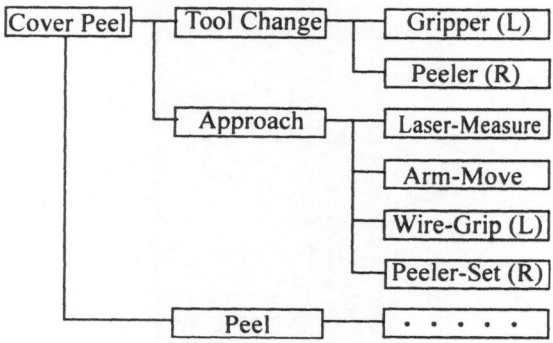

Figure 3.5   KYUDEN language

The robot language, named the KYUDEN language, has great benefits. It allows field engineers to prepare programs more quickly for individual tasks, and to evaluate work results at the work site. Software using the KYUDEN language can be made compatible with other manufacturers' programs. The language can be easily installed via floppy disks, which yield efficient management of software.

3.1.7 Automatic Control

Major new technologies in the automatic control of Phase II are listed below:
- Dual-arm coordination
  In order to operate dual arms in a safe and efficient manner, the following coordination controls are introduced in Phase II.
  - Follow-up motion control
    Recall that some heavy objects must be manipulated by two arms. Follow-up

motion control helps an operator to maneuver the two arms by controlling a single arm. The operator controls a single arm while the other arm follows the motion of the controlled arm so that two arms can perform tasks coordinately. The high accuracy of the arm motion has enabled this function.

• Work sharing between two arms
  Note that many operations in live-line maintenance require dual-arm coordination, or can be performed easily using two arms. For example, insertion of a sleeve needs dual-arm coordination. The left arm holds a wire and the right arm grasps a sleeve and inserts the wire into the sleeve. The motion of the right hand is determined according to the position and the orientation of the wire, which is held by the left hand. (Figure 3.6)

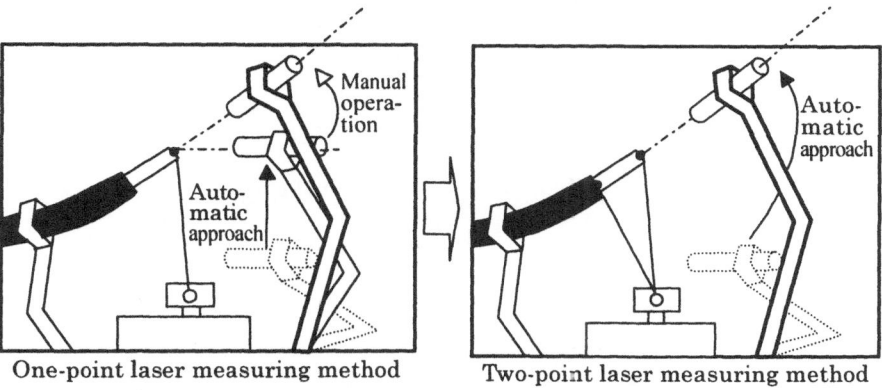

One-point laser measuring method   Two-point laser measuring method

Figure 3.6   Dual-armed operation for sleeve-insertion via two-point measurement

• Prevention of mutual interference between dual arms
  To prevent mutual interference, the position and the orientation of dual arms are controlled to avoid interference between the arms and to improve work efficiency.

• Impedance control
  In order to realize compliant robot motion, an impedance control technique has been introduced. Compliant motion can be realized by virtual joint impedance. In the wire tension puller operation, for example, the left arm folds the tension puller and sets it on a wire. The right arm operates the tension puller. The wire is tightened and moved upwards as the puller applies tension on the wire. Then both arms rise as the puller moves upward. It is not necessary to move the robot body upwards during the operation since arms follow the motion of the tensioned wire. This function has made the task for the operator easy.

- Human-machine interface

  A touch-panel display was developed to simplify the communication between an operator and the robot. (Figure 3.7) By touching multiple panels installed inside the operating cabin, the operator sends commands to the robot. The number of control switches was reduced to a minimum after the development of the touch-panel display. In addition, computer graphics were introduced to show the locations of the robot and the spatial relationship between the robot and work objects. This function can support the monitoring of the working environment and the auxiliary operation during the maintenance.

Figure 3.7   Human-machine interface in the cabin

3.1.8 Electrical Insulation Technology

Unlike general industrial robots, a LLW robot requires a high standard of electrical insulation. Insulation on Phase II is effective both on the robot arm and the boom of the vehicle. This allows 6kV/22kV LLW to be carried out even in rainy conditions. Insulation on the robot arm is installed at the forearm (wrist and elbow). The driving unit for the wrist, which has three axes, is installed at its elbow. Concentric insulated pipes are housed in the forearm. These pipes transmit driving power to the wrist. In addition, an insulated cover is installed on the outer surface of the forearm to prevent discharging in rainy conditions. A cover is also installed at the farthest end of the boom, in addition to the insulated section of the boom end. This modification helps secure insulation tolerability and allows for the safe maintenance of 22KV systems in rainy conditions. This method was found to be more reliable and require less maintenance than the compound coating method, which had been undergoing simultaneous development. (Figures 3.8 and 3.9)

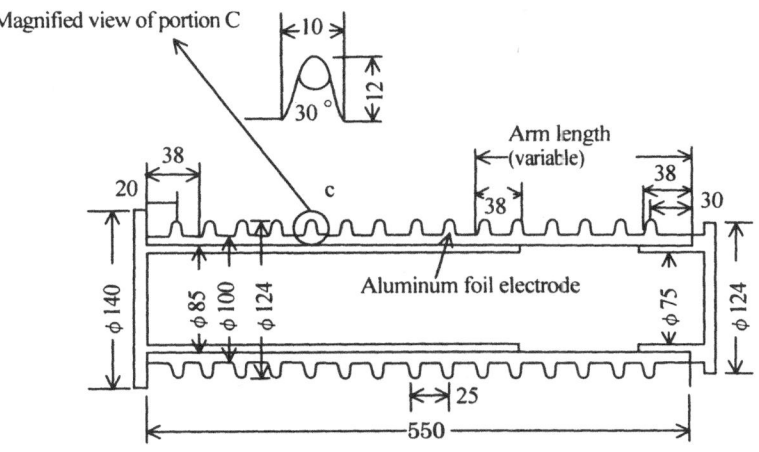

Figure 3.8   Simulated insulated arm

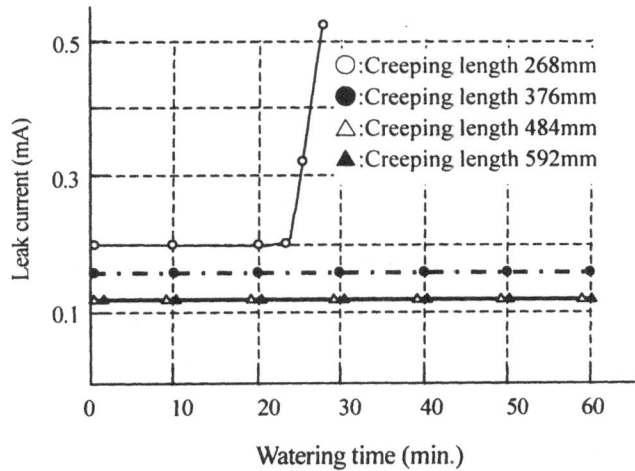

Figure 3.9   Characteristics of leak current of simulated insulated arm

3.2 Application to Field Work

Phase I robots, which were already in use at all 85 customer service offices by 1992, are widely used throughout the field for 90% of company maintenance tasks. This indicates the suitability of the robot as a LLW tool. Development of Phase II was completed in December, 1997. A field test of 17 Phase II units started at the end of 1999 after testing and operator training. (Figure 3.10) A demonstration of new distribution techniques was held in June, 1998 by Kyushu Electric. About 400 personnel from Japan and guests from overseas attended the demonstration. Kyushu Electric's technology and the indirect LLW technique were compared with each other through the demonstration.

Figure 3.10   Switchgear installation using Phase II

The replacement of 6kV insulators was performed simultaneously using three different methods: Phase I, Phase II, and hot-sticks. The time required for the three methods was compared. Phase II was found to require two workers (1 operator and 1 supervisor), Phase I needed 3 workers (1 bucket operator, 1 temporary support operator and 1 supervisor), and the hot-stick method required 4 workers (2 bucket operators, 1 tool delivery assistant, and 1 supervisor). Even though the working time is nearly the same, the use of robot technology requires fewer workers and the cost can be reduced. In the demonstration, it was found that a worker using the hot-stick method can start maintenance operations immediately, while an operator using the Phase II method cannot start right away since time is needed for preparing the complicated laser-based measurement system. However, the demonstration eventually revealed the superior performance of Phase II, which took 44 minutes to complete the operation. It took 48 minutes by the hot-stick method and 55 minutes by Phase I. The hot-stick method required 3.2 man-hours to complete the operation. Phase I required 2.75 man-hours (14% less). Phase II required 1.47 (or 54% less) man-hours. With a few modifications, Phase II is expected to perform operations quicker in the near future. This will result in the reduction of maintenance time. (Table 3.1)

Table 3.1   LLW comparison in man-hours for changing 3 insulators

Phase II	Hot-sticks	Phase I	Phase II
Time (min.) (a)	48	55	44
Workers (person) (b)	4	3	2
MH (a)×(b)	3.20	2.75	1 .47
Comparison	Base	-0.45 (-14%)	-1.73 (-54%)

4. Evaluation of LLW Robots

4.1 Benefits of Outage-free Maintenance Technique

As outlined at the beginning of this paper, LLW robot development was initiated as a technique for outage-free maintenance. Extensive deployment of outage-free maintenance, not including robot techniques, has yielded cost reductions beyond expectations. These savings have been put back into the research and development of robot technology to accelerate its progress. (Figure 3.11)

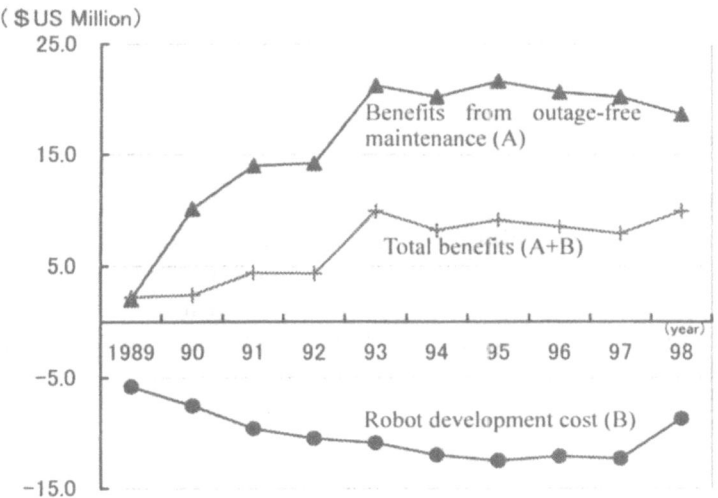

Figure 3.11   Investment into robot development using
savings from outage-free maintenance

Since tackling the improvement of outage-free maintenance techniques in 1984, maintenance outage on medium voltage lines was eliminated altogether in 1988, six years earlier than expected. In 1989, maintenance outage on low voltage lines was also eliminated, not only improving customer service but also benefiting power companies and maintenance companies. It had been generally believed that expensive, heavy equipment for maintenance or installation, such as bypass cable trucks, mobile transformers and mobile generators, would increase working costs more than outage work does. However, a cost analysis revealed that the total expenses involved in outage, such as notifying customers of power interruptions and paying workers holiday or night wages, are higher than those of installing equipment.

4.2 Needs for and Anticipated Benefits of Robot Technology

The birthrate in Japan has decreased from 2.09 million births in 1973 to 1.19 million in 1997, a decline of 40%. This trend decreases the labor force. Given this situation, further labor savings will be urged in the power distribution sector using mechanized force. Many benefits are anticipated from the application of robot technology to distribution line work considering the increasing demands for a high quality power supply at a low cost. Some expected benefits include the following:

- Realization of cost reduction through enhanced work efficiency achieved by fewer workers
- Creation of a better work environment during rainy, hot, and cold conditions by mechanizing overhead LLW
- Enhancement of the reliability in power facilities by eliminating inconsistent work skills, and by maintaining high work quality

5. Conclusion

A strong base has been established to develop ongoing robot technology. Sophisticated, high quality distribution line work can be performed at a low cost in a safe, comfortable working environment. Kyushu Electric believes this is a key in satisfying the demands of the society of the future in which there will be acute labor shortages. Wider field application of LLW robots will yield innovative distribution work in the 21st century and will realize further cost reductions.

Reference List

Journal Articles
1. Maruyama Y, Morita K 1994 Mobile robot work hot lines. *Transmission & Distribution International* Vol. 5 No 3 Third Quarter: 10-18
2. Nakashima M, Harada S, Yano K, et al. 1996 Development of a robot language for hot-line work. *Advanced Robotics* Vol. 10 No 4

Papers in Proceedings
3. Maruyama Y, Hizen Y, Yakabe H, et al. 1994 Development of hot-line work robot. *2nd International Conference on Live Maintenance (ICoLIM'94)*. Mulhouse, France
4. Yakabe H, Maruyama Y 1995 Elimination of maintenance outage and cost reduction by development of outage-free maintenance techniques. *IEA-UNIPEDE New Electricity '21*. Paris, France
5. Maruyama Y, Yano K, Nakashima M 1995 Application of semi-automatic robot technology on hot-line maintenance work. *Distribution 2000*. Brisbane, Australia
6. Tanaka S, Maruyama Y, Yano K, et al. 1996 Work automation with the hot-line work robot system Phase II. *International Conference on Robotics and Automation (ICRA)*. Minneapolis, U.S.A.
7. Maruyama Y, Yano K, Matsuo T, et al. 1996 Development of a distribution hot-line work robot and a robot simulator. *International Conference on Electrical Engineering (ICEE '96)*. Beijing, People's Republic of China
8. Someya Y, Maruyama Y, Yano K, et al. 1997 Study on performance of 22KV all-weather insulation for hot-line robot. *International Conference on Advanced Intelligent Mechatronics '97*. Tokyo, Japan
9. Tsunemi K, Maruyama Y, Yano K, et al. 1998 Development of a hot-line work robot Phase II and a training system for robot operators. *ESMO '98*. Florida, U.S.A., pp 147-153
10. Yano K, Maruyama Y, Nakashima M 1998 Development of the live-line work robot Phase II for MV overhead lines. *4th International Conference on Live Maintenance (ICoLIM '98)*. Lisbon, Spain, pp 459-464

Section 5.4

Flexible Automatic Wiring of Long-Tube Lighting and Service Cabinet Modules

K.-F. Kämper

1. Introduction

It is well known that the automatic laying of compact and stranded wires involves considerable difficulties. Nevertheless, the lighting industry insisted on the development of a flexible automatic cabling unit for the wiring of long-tube lighting with plug components - with the goal of reducing labor costs, of simplifying production procedures (previously, the wiring had to be prefabricated), and of uninterrupted line production (previously, the production processes were operated with in-process stocks). The long-term goal is to link together the punching and folding of the sheet metal parts, the enamelling, the mounting of the components, the wiring, and the testing and packaging of the entire lighting onto a single line and to achieve an improvement in quality.

Similar conditions hold for the flexible automatic wiring of service cabinet modules, namely reduction of labor costs, simplification of organization and improvement of quality.

Systems have previously been introduced onto the market which

1. use a robot to take up prefabricated wires and plug them into pre-positioned components
2. position wires and plug them into cut-and-clamp components (IDC components) (Disadvantage: the previous plug components cannot be altered during the sequence)

A wiring system has now been developed which utilizes 6-axis industrial robots for flexible wiring of long-tube lighting with plug components, providing a high degree of flexibility for the different lighting types and the positions of components. The system extracts in sequence the wires to be positioned. The section thus extracted is then transported to the placement head and cut there through the middle, so that both ends can be used, in order that the open wire end can then be plugged into the components through the use of the placement head (see Figure 1).

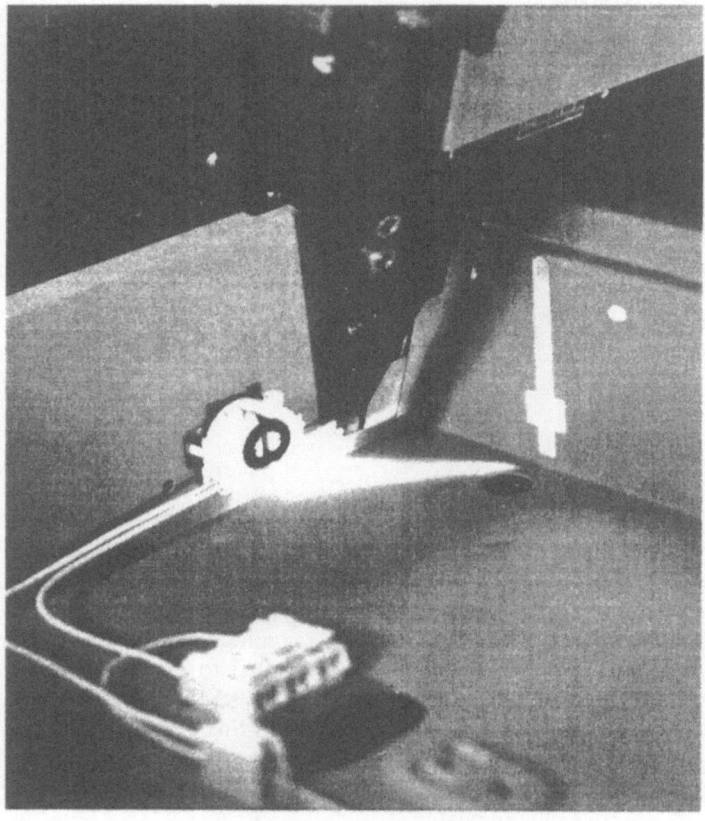

Figure 1: Placement head

After this system was introduced onto the market for the lighting industry, the components industry also began to call for a system based upon the technology described above, to be used for the wiring of service cabinets. A system for the wiring of service cabinets and the prefabrication of service cabinet components was developed for this purpose. With this the strands are separated from one another in advance, the free ends made compact, the cable ends marked with lettering (or, where needed, with colors, in stripes or solids), coded and then, with the aid of the robot, flexibly and automatically wired and placed in the corresponding components using the cage clamp technique. The prerequisite here is that not only such components as clamp bars and plug connectors be equipped with cage clamp technique components, but others as well, including relays, etc.

Both systems have since found application in everyday practice. The technology has been patented.

2. Automatic flexible wiring of lighting

The requirements of the lighting industry can be recapitulated as follows:

1. Traditional plug components such as lighting brackets, capacitor brackets, connection terminals, and fluorescent lamp ballasts (both electronic and traditional) are to be flexibly wired automatically.
2. No significant redesigning of the lighting construction is to take place as a result of the introduction of the automatic wiring system.
3. The installation is to be economically practical.
4. The installation is to provide a high degree of flexibility in terms of different lighting constructions.
5. The installation is to be easily operated, easily programmed and user-friendly.
6. It is to be usable with the entire range of lighting sizes, including 4 x 18 W (600 x 600 mm), 4 x 36 W (600 x 1200 mm), 2 x 58 W (300 x 1500 mm) and 2 x 70 W (300 x 1700 mm), as well as with various other lamps, such as T8s and T5s.
7. It must be able to function with plug components from a variety of suppliers. The components must be able to be plugged in either horizontally, vertically or at an angle.

2.1 Description of the installation

Figure 2: Installation setup

The installation consists of (see Figure 2):

- a 6-axis industrial robot
- an extraction device
- three traction groups
- a placement head
- a retaining table for the lighting
- safety screen
- software

2.2 Function description

A 6-axis industrial robot, with three additional traction groups that it operates simultaneously, using an extraction machine, pulls wire from out of a vessel in lengths preset by the wiring program. The wire is doubly extracted in the extraction machine, without being separated. The extracted section is transported through various buffers up to the placement head of the robot, bent into a horizontal position, cut there in the middle, so that the first end becomes free and the robot can plug the first wire end into the first component. After attaching the open wire end in the component clamp, the robot transports the wire at path speed further to another component. There the next extracted section is separated. The other end is gripped by the gripper. The free end of the head is raised vertically so that the gripper can plug the free end into the next component. This procedure is repeated until the lighting unit is completely wired.

Industrial robot

A 6-axis robot with an extensively-equipped controller with several PC boards was chosen so that components in the lighting space could also be reached. The robot is additionally equipped with three axis reinforcers for the operation of three traction groups, which are operated simultaneously by the robot computer. The robot can be equipped with an image processing system, which can process any image data for correction in real time via a system bus. Devices called force exertion sensors can detect the wire's traction and pressure forces, by which quality monitoring is carried out. The additional traction groups operate a traction group once after the extraction device. Here the different wire lengths are intermittently first-drawn via robot operation, in accordance with the selected program, and then extracted in the commercially-available extraction device. The wire travels through a storage collector to the next traction group, so that the wire can be transported there in a stress-free state to the next traction group in the robot gripper (see Figure 3).

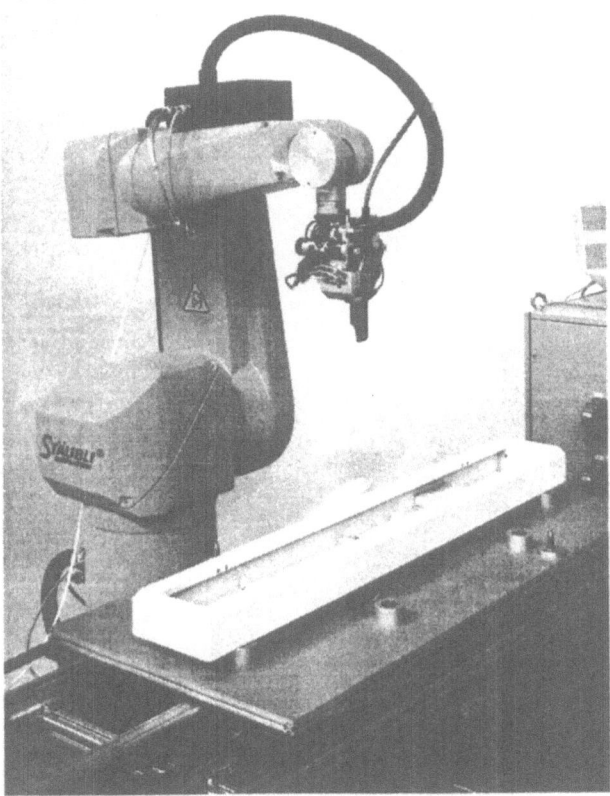

Figure 3: Industrial robot equipped with placement head

Placement head

Except for the software, the placement head is the most important component on the robot. This is where the wire is precisely transported and directed through the use of a guide pin. The midpoint of the wire is located with additional sensors and encoders in real time, so that the cutting blade in the placement head can separate the wire in the middle. The placement head has an additional gripper for the other end of the wire and a Z-movement for the pin, which clears a path for the second insertion through an upward movement of the pin. Additional cylinders ensure that the wire is also bent, so that vertical insertions can be carried out. Certain laser sensors correct the wire at the midpoint through the corresponding encoder. The quality of the placement head provides certain antifriction characteristics, in order that the wire can be guided by the pin with as little resistance as possible. The placement head must also be so constructed that relatively high lighting units can be wired, which means that the corresponding pin must present a certain length. The 6-axis robot also makes it possible to wire components which are mounted laterally or in concealed positions - as well as such things as those lampholders, for example, which are located in concealed positions in plastic end boxes (see Figure 4).

Figure 4: Placement head in detail

Retaining table for the lighting units

The table is equipped with permanent magnets for putting the lighting unit firmly in place. Interchangeable pins provide take-up and exact positioning of the lighting units.

A protective screen in accordance with the safety regulations ensures the necessary safety. Alternately, the tables themselves can be made with carriage-type construction (according to customer preference). The entire cell can also however be used in conjunction with a linear transfer system with rail heads.

Software

The software contains movement parameters for the robot head and/or arm, the positions of the individual components and additionally the wire advance drive modification for the various lengths of the wires to be positioned. Furthermore, the robot operation, which is programmed on a multitask basis, operates the different sensors and the inputs and outputs of the entire system, eliminating the need for a separate SPS. Certain algorithms relating to the components and the procedural strategy in curves and approach strategies for components are present in the system in parameter form. Besides this, the system continuously compares the lengths of the wires, using the resolvers which are attached to the different traction groups. The teaching of the system can be carried out manually after compilation of the wire placement plan for the lighting. Corrections can be entered at the terminal or the entire system can be impinged upon using a special CAD simulation system. The software graphics are designed in such a way that the operator is led through the program and is informed of errors through plain writing displays. In addition, statistics are possible for better operating data preparation.

2.3 Additional technical requirements

Under the influence of increasingly smaller batch sizes and increasingly large product variety, the lighting industry is promoting a rapid reprogramming towards the corresponding lighting types. With this goal in mind, a CAD robot simulation system based on this principle was modified in such a way that it could be applied to the lighting industry. Thus, with this simulation system, existing 3-D lighting constructions could be transferred from customer computers via corresponding interfaces. With the help of the mouse, the design engineer places the corresponding components, such as lighting bracket, ballast or terminal block, within the lighting unit, based upon the CAD data of the components. In this way, he complies a wiring plan based on the expert system. With the aid of this construction and of the positioning plan, the program can be simulated on the PC, in order to test any possible crash situations. After corrections, the entire program can be loaded onto the robot. Because the robot is specially calibrated, the data can be implemented almost without correction directly into the robot. This means that the robot can begin the wiring at once (see Figure 5).

Figure 5: ADS – Automatic Direct wiring of Standard components

2.4 Results, experience, performance

Already more than 20 installations have been put into operation in Europe, in production line form as a rule with integrated railhead transport and automatic testing systems and with the option of subsequently automating manual labor positions. The installations operate with high levels of service quality.

In order to install and operate this kind of computerized installation, is absolutely necessary that the user's employees, operators and maintenance personnel all be trained accordingly.

Statements regarding cycle times can as a general rule not be simplified, because every lighting type is individually designed. One can however say that 4 x 18 W lighting requires a cycle time of approximately 60 seconds while 2 x 36 W lighting, as a comparison, needs approximately 40 seconds.

For wiring of service cabinets, one must base one's calculations upon approx. 6.5 to 7 seconds per wire and two insertion points, of which a considerable time passes for the opening of the two cable cage clamps. The length of the cable has no decisive role to play in this regard.

2.5 Future development

Uninterrupted line production

The lighting industry regards the investment in wiring installations as a welcome opportunity to construct uninterrupted line production from the processing of sheet metal through the coloring unit all the way to the packaging process. Previously, it was chiefly wiring installations with integrated rail head transport systems and integrated automatic, flexible testing installations that were constructed and supplied for this purpose. The testing installation inspects the wired lighting on the rail head (which is supplied by the transport system of the <u>testing installation</u>) for <u>functional, insulation, grounding</u> and <u>high voltage testing.</u> The testing installation can be programmed, and the parameters can be entered in accordance with the regulations. Attached label printers make it possible to designate the tested lighting accordingly and thus improve the quality image of the lighting manufacturer. Faulty units are marked with the respective faults for better repairing. Documentation and statistical evaluation at the testing installation computer are tools for further improvement of lighting quality (see Figure 6).

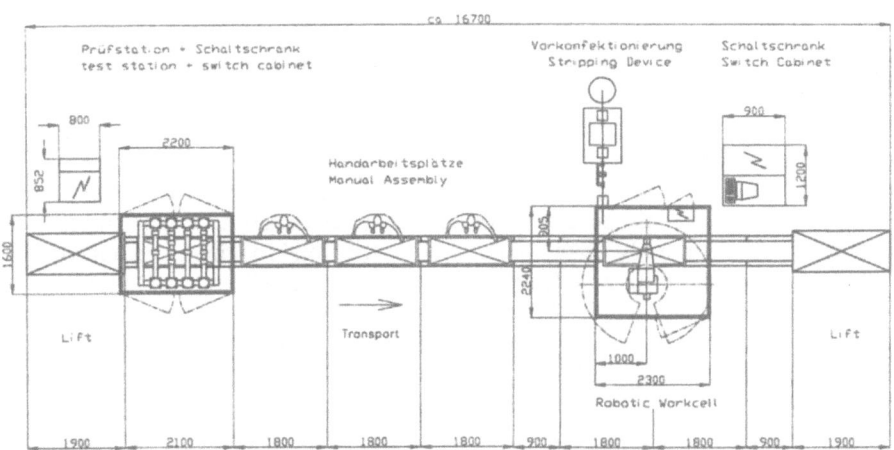

Figure 6: Layout of ADS system

Modular construction

Manual work stations on the transfer system are set up for component pre-assembly. The system is modularly so constructed that these underlined manual work positions can be replaced at any stage by flexible automated work positions. In the long run, the trend is toward fully automated lines from sheet metal processing through flexible automatic placement of the lighting units, wiring and testing all the way to the packaging stage.

Teleservice

The installations are sold worldwide, as a result of their durability. In order to improve after sales service, teleservice will have an important role to play in the future, e.g. long-distance diagnostics via PC, alterations of the program being played back into the robot by the supplier at the customer location and on-site video conferences at the installation for the purpose of solving concrete mechanical problems.

Stranded wire and coloring

Certain markets demand stranded wire, which is to be made compact through the use of ultrasound or colored according to the program. There are solutions for both systems that ensure that the wire can be plugged in. Multicolor inkjet systems make it possible, depending on program selection, to have the colors appear on the wire as solids, stripes or rings. A printed barcode facilitates monitoring of the entire process.

2.6 Conclusion for automatic flexible systems wiring of lighting

With the development of flexible automatic wiring of lighting units, an instrument has been put into the hands of the lighting industry for the first time which, in addition to highly-automated sheet metal processing, will allow the automation of the greater part of the other production activities, such as pre-assembly of the lighting units, wiring, testing, packaging, etc. The majority of the users utilize the wiring installation to change over their lighting manufacture to uninterrupted line production. This is only made possible by the total computerization of the installations and the easy adjustability and user-friendliness of the overall installations. The setting up of this installation also makes future developments (new lamp types or other new components) operable, so that these installations are also adaptable at any time through easy modification of the programs, without making a new investment mandatory.

3. Flexible automatic wiring of service cabinet mounting plates and components, such as row clamps and plug-in bars, etc.

Automation in the service cabinet industry and in the prefabrication of service cabinet elements was previously limited to the automation of the prefabrication of wires, e.g. the stripping and lettering of wires on separate fabrication machines. The wires themselves had to then be manually connected, which in addition to the organizational expenditure also meant a high manual labor cost factor for service cabinet manufacture, as well as an uncertainty regarding quality. The introduction of the cage clamp of the Wago Co. led years ago to the simplification of manual wiring in that the cage clip can be opened with a screwdriver, after which the wire can be more simply slid manually into the cage clamp. The wire was then clamped after the spring closed. Because of the high labor costs and the situation with foreign competition, the demand for automated wiring of service cabinet mounting plates with components (equipped with cage clamps) or flexible automatic prefabrication of service cabinet plug and row clamps, etc. With the help of the system developed in the lighting industry, which involves preliminary extraction of the wire according to pre-programmed lengths and subsequent separation of the wires in the placement head of the robot, a system was developed that is able to carry out the wiring of service cabinet mounting plates and/or the preliminary wiring of service cabinet components.

3.1 Requirements

The robot system must be able to wire service cabinet mounting plates up to approximately 2 x 1 m. For this it is also necessary that all of the components (such as row clamps, plug connectors, relays, transformers and clamp bars) on printed boards and similar/other parts be equipped with cage clamps which are suitable for automatic wiring. Wires with the gauge numbers 0.5 / 0.75 / 1.0 and 1.5 mm² are to be processed for automatic insertion as stranded wires. They must be lettered at their ends with the corresponding markings. Depending on the program selection, they must be capable of having colors printed on them, either in solids, stripes or rings. In addition, the system should be able to apply and wire automatic programs from the E-CAD system (through a special CAD system used as an expert system). The system is to be economical, easily operated and easily programmed.

3.2 Description of the installation

The installation consists of the following:

- a 6-axis industrial robot equipped with an integrated image processing system
- an extraction device
- an ultrasound compacting device
- a coloring and lettering installation
- three traction groups
- a placement head
- a retaining table for the service cabinet module
- safety screen
- software

The installation consists of a 6-axis industrial robot with controller which is also able to execute spatial movements in such a way that a variety of components at different heights in different positions at different angle settings can be wired. The robot is additionally equipped with an integrated image processing system in order to be able to automatically compensate for any deviations encountered while mounting the components. The robot system is equipped with additional traction groups in such a way that the transporting of the wire through the extraction, compacting and/or coloring systems described below can take place simultaneously by means of the robot operation. The traction groups, operated simultaneously by the robot, transport the wire through the coloring machine, through a storage collector to the extraction machine, then to the ultrasound compacting installation and through a storage collector to the placement head of the robot (see Figure 7 and 8).

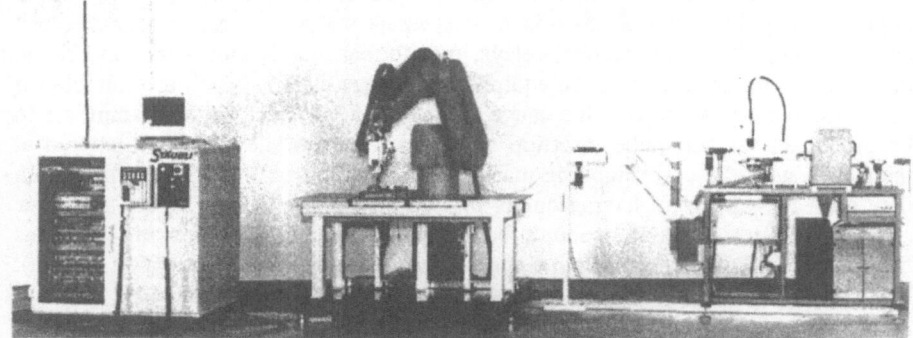

Figure 7: CC-Matic automatic wiring device for cage clamp connection technology

Figure 8: Placement head in detail

The placement head of the robot has the function of transporting the wire in the different lengths, and to be true to the dimensions while doing so, up to the point of precise separation of the extracted piece in its center. A blade is used to cut the wire at this point. Additional sensors, in conjunction with the drive and the encoders in the head, ensure a precise positioning of the wire for the purpose of dividing it. A torsion module is responsible to the rotation of the wire as second insertion. A clamping device clamps the wire for insertion of the first wire end. Together with the resolver connection, the wire drive is responsible for the precise transporting of the wire at robot speed to the components while the wire is being laid.

Extraction unit

A standardized extraction unit with adjustable tools is responsible for the extraction of the wires. The traction group belonging to it, operated by the robot, is responsible for the intermittent system first-drawing different lengths.

Color system

A color system, made up of quadruple inkjet heads for each color, ensures the lettering, coding and coloring in solids, stripes or rings of the wire, based on program selection.

Ultraschall compacting system

An ultrasound compacting system with special tools compacts the single conductors at the extracted section in such a way that no single conductors become free during insertion or separation of the wire, so that the wire can be perfectly inserted into the cage clamp.

Different storage collectors between the systems ensure the continuous running of the wiring operation, independent of the speed. Built-in coding lasers and sensors operate the storage collectors and the elements connected in series in such a way that there is always sufficient material in movement.

3.3 Function description

The robot pulls the stranded wire, in accordance with the length setting of the program, out of a vessel and through the color system, which has its own drive. Here the wires are colored in the desired lengths according to program selection (with stripes, rings or solids) and lettered and coded on the ends according to the circuit diagram. Additional drying procedures are responsible for a rapid drying of the color between the four-color inkjet heads. Using its traction group, the robot pulls the thus partially pre-colored and lettered wire through a storage collector in correct position up to the extraction machine. Extraction is carried out intermittently here. In a further intermediate step, with the help of the next traction group, the extracted section is moved to the ultrasound compactor in order to compact the single conductors into the proper form. Afterwards, using an additional storage collector and a further traction group, the wire is transported free from stresses through a special tube to the placement head of the robot. The traction group with belt drive is responsible for the pre-programmed transporting of the wire up to the extracted section. A swinging blade separates the wire in the middle at the extracted section. The wire is clamped. A pneumatically-driven mandrel moving in the lead opens the cage clamp. The wire is transported in its wake and the opened spring is introduced, the mandrel is removed, the spring closes, the wire is clamped and the robot proceeds to the next contact point while simultaneously transporting the wire using the traction group in the head. The blade separates the wire. The second end is turned by 180° and clamped. The mandrel opens the cage clamp, the wire is introduced and clamped. The process repeats itself until the mounting plate is completely wired. An image processing system automatically corrects any mounting errors.

A changing table for taking up mounting plates or service cabinet components is responsible for taking up the components and for the reciprocal equipping of these components. A safety housing provides the necessary system security while the machine is running.

Software

The software links movement programs, positions to be approached (components) and operates the entire periphery of the system, such as the extraction machine, compacting unit and coloring system.

Special parameters are responsible for the different transporting and the approach strategies of the various components. The parameterization of the software is given over to an E-CAD system. Previously, this E-CAD system permitted the designing of circuit diagrams with the input of components. This software has been expanded to include the plug position of the various components, which are taken from a library. The library has both the XYZ and the angle data of the various components, which are automatically taken into consideration. In addition to that, the system applies the restriction of the robot wiring, so that through this the entire robot program for laying wire can be generated automatically, once determination is made of placement sequence of the wire lengths, the lettering and the colors. Because this is an expert system, the system automatically generates the priorities and rejects situations where wire placement is not possible. This E-CAD simulation system was previously not available on the market and was specially developed for these matters, including the parameterization for the integrated image processing system.

Additional technical requirements

- Permanent further development of the E-CAD simulation system
- Possibility of Teleservice with error display and remote diagnostics as well as external program alteration
- Possibility of being able to reduce service costs in the future through video conferencing

In the prefabrication of service cabinet components, it should be possible to plug in looms of cables on one side only or to connect clamp bars and plug connectors together in such a way that several meters of wire are present as intermediate lengths, in order to ensure the bundling of the wires and to link everything to the CAD system.

The wiring time including the opening of the two cage clamps to be connected is approximately 6 seconds, which corresponds to approximately 1/10 of the manual wiring time.

3.4 Conclusion for flexible automatic wiring of service cabinets

With the development of this sort of a flexible automatic wiring system for service cabinets and service cabinet modules, it has become possible for the first time to automatically wire flexible conducting wires in service cabinets, while at the same time reducing the costs and improving the quality, because subsequent quality testing is a great deal easier than it is with manual wiring. The E-CAD system developed with this permits the compilation of circuit diagrams with information relating to components found in the library and the automatic generation of wiring placement plans. No prefabrication of wires on separate machines is necessary, which additionally simplifies the organization. The prerequisite for this is that a further development of components in the area of the cage clamp be continued, so that in the future all components for the service cabinet can be automatically wired.

4. References

Product names

Lighting wiring system of BJB GmbH & Co. KG, Werler Str. 1, 59755 Arnsberg:
ADS Automatic **D**irect wiring of **S**tandard components
Wiring system for service cabinet modules from Wago Kontatktechnik, Hansastr. 27, 32423 Minden:
CC-Matic automatic wiring device for cage clamp connection technology

Patents

Protective Right No. for automatic wiring of long-tube lights: 95 14 629
Protective Right No. for automatic wiring of service cabinet modules: 97 56 750